THE OIL SPILL RECOVERY INSTITUTE

PAST, PRESENT, AND FUTURE DIRECTIONS

Committee to Review the Oil Spill Recovery Institute's Research Program

Polar Research Board

Ocean Studies Board

NATIONAL RESEARCH COUNCIL
OF THE NATIONAL ACADEMIES

THE NATIONAL ACADEMIES PRESS
Washington, D.C.
www.nap.edu

THE NATIONAL ACADEMIES PRESS • 500 Fifth Street, N.W. • Washington, DC 20001

NOTICE: The project that is the subject of this report was approved by the Governing Board of the National Research Council, whose members are drawn from the councils of the National Academy of Sciences, the National Academy of Engineering, and the Institute of Medicine. The members of the committee responsible for the report were chosen for their special competences and with regard for appropriate balance.

This study was supported by Contract/Grant No. 01-01-23 between the National Academy of Sciences and the Oil Spill Recovery Institute. Any opinions, findings, conclusions, or recommendations expressed in this publication are those of the author(s) and do not necessarily reflect the views of the organizations or agencies that provided support for the project.

International Standard Book Number: 0-309-08514-4

Copies of this report are available from:

Polar Research Board
2101 Constitution Avenue, N.W.
Washington, DC 20418
(202) 334-3479

Additional copies of this report are available from the National Academies Press, 500 Fifth Street, N.W., Lockbox 285, Washington, DC 20055; (800) 624-6242 or (202) 334-3313 (in the Washington metropolitan area); Internet <http://www.nap.edu>.

Cover: Photo of the Prince William Sound in Alaska. Photo courtesy of the Exxon Valdez Oil Spill Trustee Council. Design by Michele de la Menardiere of the National Academies Press.

THE NATIONAL ACADEMIES
Advisers to the Nation on Science, Engineering, and Medicine

The **National Academy of Sciences** is a private, nonprofit, self-perpetuating society of distinguished scholars engaged in scientific and engineering research, dedicated to the furtherance of science and technology and to their use for the general welfare. Upon the authority of the charter granted to it by the Congress in 1863, the Academy has a mandate that requires it to advise the federal government on scientific and technical matters. Dr. Bruce M. Alberts is president of the National Academy of Sciences.

The **National Academy of Engineering** was established in 1964, under the charter of the National Academy of Sciences, as a parallel organization of outstanding engineers. It is autonomous in its administration and in the selection of its members, sharing with the National Academy of Sciences the responsibility for advising the federal government. The National Academy of Engineering also sponsors engineering programs aimed at meeting national needs, encourages education and research, and recognizes the superior achievements of engineers. Dr. Wm. A. Wulf is president of the National Academy of Engineering.

The **Institute of Medicine** was established in 1970 by the National Academy of Sciences to secure the services of eminent members of appropriate professions in the examination of policy matters pertaining to the health of the public. The Institute acts under the responsibility given to the National Academy of Sciences by its congressional charter to be an adviser to the federal government and, upon its own initiative, to identify issues of medical care, research, and education. Dr. Harvey V. Fineberg is president of the Institute of Medicine.

The **National Research Council** was organized by the National Academy of Sciences in 1916 to associate the broad community of science and technology with the Academy's purposes of furthering knowledge and advising the federal government. Functioning in accordance with general policies determined by the Academy, the Council has become the principal operating agency of both the National Academy of Sciences and the National Academy of Engineering in providing services to the government, the public, and the scientific and engineering communities. The Council is administered jointly by both Academies and the Institute of Medicine. Dr. Bruce M. Alberts and Dr. Wm. A. Wulf are chair and vice chair, respectively, of the National Research Council.

www.national-academies.org

COMMITTEE TO REVIEW THE OIL SPILL RECOVERY INSTITUTE'S RESEARCH PROGRAM

MAHLON C. KENNICUTT, II, *Chair,* Texas A&M University, College Station
BRENDA BALLACHEY, U.S. Geological Survey, Anchorage, Alaska
JOAN BRADDOCK, University of Alaska, Fairbanks
AKHIL DATTA-GUPTA, Texas A&M University, College Station
DEBORAH FRENCH MCCAY, Applied Science Associates, Naragansett, Rhode Island
JERRY NEFF, Battelle Memorial Institute, Duxbury, Massachusetts
JAMES PAYNE, Payne Environmental Consultants, Inc., Encinitas, California
JAMES RAY, Shell Global Solutions, Inc., Houston, Texas
WILLIAM SACKINGER, OBELISK Hydrocarbons, Ltd., Fairbanks, Alaska

Staff

CHRIS ELFRING, Director, Polar Research Board
DAN WALKER, Senior Staff Officer, Ocean Studies Board
ANN CARLISLE, Administrative Associate

RALPH S. LEWIS, Connecticut Geological Survey, Hartford
BONNIE MCCAY, Rutgers University, New Brunswick, New Jersey
JULIAN P. McCREARY, JR., University of Hawaii, Honolulu
JACQUELINE MICHEL, Research Planning, Inc., Columbus, South
  Carolina
RAM MOHAN, Gahagan & Bryant Associates, Inc., Baltimore,
  Maryland
SCOTT NIXON, University of Rhode Island, Narragansett
JON SUTINEN, University of Rhode Island, Kingston
NANCY TARGETT, University of Delaware, Lewes
PAUL TOBIN, Xtria, Chantilly, Virginia

Preface

The Oil Spill Recovery Institute (OSRI) is a small organization based in Cordova, Alaska, but it has a large task. Established as part of the Oil Spill Prevention Act of 1990, it was charged to identify and develop methods to deal with oil spills in Arctic and subarctic environments and work to better understand the long-range effects of oil spills on the natural resources of Prince William Sound and its adjacent waters, including the environment, economy, and people. It is a small program, disbursing about $1 million each year. But it works in a critical and challenging area: helping the nation prepare for oil spills in cold regions.

The committee conducted this review much like a visiting committee review of a university program. Our nine members traveled to Alaska in February 2002 to gain an understanding of the program, its mission, and its research and technology projects. We talked frankly with the Advisory Board, scientists, and staff about the accomplishments and challenges of the program. We distributed a call for input by email and received comments from others who knew the program, and this information, although anecdotal, gave the committee broad insights into how the community perceives the OSRI program and helped us formulate our conclusions. We reviewed as many documents as we could in the time available to us: the Grant Policy Manual, sample calls for proposals, sample proposals, meeting minutes, and many other reports. Finally, we sent numerous sets of questions to the OSRI staff as we dug deeper into the program, to be sure that we understood what they did and how they did it. We then held two writing meetings, where we reviewed materials, deliberated, and wrote the final report. This report is not intended to be a project-by-project review of OSRI activities, but instead is a broad assessment of the program's strengths and weaknesses, with special emphasis on whether the activities supported are addressing the OSRI mission, whether the processes used are sound, and whether the research and technology projects are of high quality.

Many people provided important information to our committee as we prepared this report. In particular, the committee would like to thank Dr. Gary Thomas, Director of the Oil Spill Recovery Institute, for his insights. Special thanks go to Ms. Nancy Bird and the other OSRI staff for their diligence and patience in responding to our requests for information. We also want to thank Dr. John Calder, current chair of the OSRI Advisory Board, for his leadership and all of the members of the Advisory Board and the Scientific and Technical Committee for their input.

On behalf of the entire committee, I want to express our appreciation to the Polar Research Board's staff, Chris Elfring and Ann Carlisle, and Dan Walker from the Ocean Studies Board. Their guidance kept us on track. Finally, let me add a word of thanks to the committee's members. This was a talented and thoughtful group, and they showed an exceptional ability to work together as a team. We hope that our report and recommendations provide the guidance requested as OSRI moves into its second five years of existence.

MAHLON (CHUCK) KENNICUTT, II, *Chair*
Committee to Review the Oil Spill Recovery
Institute's Research Program

Acknowledgments

This report has been reviewed in draft form by individuals chosen for their diverse perspectives and technical expertise, in accordance with procedures approved by the NRC's Report Review Committee. The purpose of this independent review is to provide candid and critical comments that will assist the institution in making its published report as sound as possible and to ensure that the report meets institutional standards for objectivity, evidence, and responsiveness to the study charge. The review comments and draft manuscript remain confidential to protect the integrity of the deliberative process. We wish to thank the following individuals for their review of this report:

Karl Turekian, Yale University, New Haven, Connecticut
Amanda Lynch, University of Colorado, Boulder
Edward Brown, University of Northern Iowa, Cedar Falls
Judy McDowell, Woods Hole Oceanographic Institution, Massachusetts
Mike Castellini, University of Alaska, Fairbanks
Cort Cooper, Chevron Petroleum Technology, San Ramone, California
Merv Fingas, Environment Canada, Ottawa, Ontario
Terri Paluszkiewicz, National Science Foundation, Arlington, Virginia

Although the reviewers listed above have provided many constructive comments and suggestions, they were not asked to endorse the conclusions or recommendations nor did they see the final draft of the report before its release. The review of this report was overseen by Dr. Garry Brewer of Yale University. Appointed by the National Research Council, he was responsible for making certain that an independent examination of this report was carried out in accordance with institutional procedures and that all review comments were carefully considered. Responsibility for the final content of this report rests entirely with the authoring committee and the institution.

Contents

APPENDIXES

Summary

I n 1989, the T/V *Exxon Valdez* ran aground in Prince William Sound and spilled approximately 11 million gallons of oil. This incident awakened the nation to how ill-prepared we were to deal with oil spills in general and especially in cold environments. There was little information available on the affected marine environment and the invertebrate, fish, and wildlife resources in the Sound. Yet information about existing resources is critical to guide decisions about effective spill response and understand long-term impacts.

As one of numerous reactions to the spill, Congress passed the Oil Pollution Act of 1990 (OPA 90). Within the legislation was a mandate to create the Oil Spill Recovery Institute (OSRI). OSRI was established to serve as a research and technology development organization, charged to provide funding to support oil-spill related research, education, and technology development projects for dealing with oil spills in Arctic and subarctic marine environments. The legislation directs OSRI to improve our understanding of the long-term effects of oil spills on the natural resources of Prince William Sound (PWS) and its adjacent waters, including the environment, the economy, and the lifestyle and well-being of the people who are dependent on them (Title V, Section 5001, Oil Pollution Act of 1990).

OPA 90 called for OSRI to be housed with the Prince William Sound Science Center (PWSSC) in Cordova, Alaska, and instructed the two organizations to share administrative functions. Funding for OSRI was initiated in 1996 and the first research funds ($200,000) were awarded in FY98. Now, OSRI has a working budget of about $1.2 million per year, generated as interest on a $22.5 million trust held by the U.S. Treasury (U.S. Coast Guard Appropriation Bill, 1996). To date, the OSRI research program has supported about $5 million of projects over the approximately five years of operation since it began making awards.[1]

The current OSRI research and technology development grant program focuses on three areas:

- **Applied Technology**—to conduct research and development on new technologies for preventing and responding to oil spills in the Arctic and subarctic;
- **Predictive Ecology**—to develop new capabilities to predict changes in populations at risk from spills; and
- **Education and Outreach**—to make the research process interactive with the public and in general provide public information and education about oil spill impacts and response.

The OSRI Advisory Board has directed OSRI administrators to strive for a 40/40/20 split of funds among these three main program areas. OSRI funding is authorized for ten years ending in 2006, and discussions about whether it will continue are beginning.

Although OSRI's research activities are still relatively new, the upcoming decision about its future makes this an excellent time for an independent review of the program. This report is not intended to be a project-by-project review of OSRI activities, but instead is a broad assessment of the program's strengths and weaknesses, with special emphasis on whether the activities supported are addressing the OSRI mission, whether the processes used are sound, and whether the research and technology projects are of high quality.

EVALUATION

Because OSRI began granting awards in FY98, this review examines a relatively brief record (FY98-FY01). In its first five years, OSRI has produced some good results and it has had some problems. Some of the problems are to be expected during the start-up phase of any new organization, such as unevenness in project quality and selection of projects somewhat peripheral to the mission. There has been a positive and ongoing evolution in policy and procedures, as experience is gained.

---

[1]Because this review focuses on the OSRI research program, it deals almost exclusively with the period 1998 to present. Some sense of OSRI's evolution and activities prior to 1998 is contained in Appendix C.

The committee believes that OSRI has supported a range of good research projects and it has the potential to become a solid (albeit small) contributor to the quest for understanding cold water ecosystems, oil spills, and their interactions. There is excellent local support and involvement, and many unanswered questions could benefit from further study. However, to be effective over the next decade, OSRI needs a new phase of strategic planning, specifically to refocus activities so they are more closely aligned with its mission. To date, OSRI has focused heavily on its modeling program. Yet OSRI could be doing much more to add to our understanding of oil and its effects on marine ecosystems, an area clearly within the OSRI mission but underserved by current programmatic emphasis. For instance, there is much to be learned about the effects of shoreline cleaners and dispersants, biodegradation, chronic effects of contaminants on nearshore flora and fauna, and long-term damage assessments. Within its technology program, OSRI is better suited for broader studies that ask "what are the best approaches for preventing and mitigating oil impacts in Arctic and subarctic environments?" rather than trying to participate in the design of specific technologies, which is very expensive and probably beyond OSRI's capabilities and resources.

Regarding OSRI's modeling activities, the committee is concerned about the disproportionate degree of emphasis given to model development and the strong operational focus. OSRI has limited resources and it simply cannot function effectively in a "real-time" oil spill response setting. Also, because model development is expensive, making this the primary focus will keep OSRI from supporting many other important activities.

Even more important is the fundamental question of whether OSRI should have a real-time operational role in responding to oil spills. A significant part of the Nowcast/Forecast (NC/FC) model is intended to serve as a tool used on a daily basis that would position OSRI to be of assistance in spill trajectory prediction and response. The committee concludes that it is not appropriate for OSRI to be involved in real-time oil spill response. Instead of striving for a real-time trajectory model, OSRI could fund research to develop modeling tools useful to address research questions and explore possible scenarios and responses. This approach to modeling could help people understand the system and its functions and forcings, and think about "what if." If developed with this goal, the model would be most useful to planners, not responders.

The size and importance of the Nowcast/Forecast effort raises a related issue: the mandated 40/40/20 split among Applied Technology, Predictive Ecology, and Education/Outreach components of the OSRI program, a requirement created by the Advisory Board to build balance into the program. This was a valid attempt to steer a new program but it has not worked as intended. It has led to some arbitrary decisions about how portions of projects are categorized and recorded. The Advisory Board should revisit this allocation mandate. They should develop a long-term strategic plan that directs the program and assures that activities support the mission.

The Predictive Ecology and Applied Technology programs are both generally responsive to the OSRI mission. Within each program, there are a few examples of activities that are not clearly directed to the mission, but overall relevance is good. OSRI's overall portfolio appears fragmented because projects are not linked by any themes or hypotheses that tie the pieces together in support of the mission. The Advisory Board should lead a strategic planning effort to clarify the OSRI mission and how it will be addressed, with concrete milestones to guide the program.

OSRI has done limited work outside of Prince William Sound, especially in the Arctic, and while the committee is fully aware that there are many needs and unanswered questions related to oil in truly Arctic marine settings, we understand OSRI's decision to focus on Prince William Sound, Cook Inlet, and the northern Gulf of Alaska: OSRI is a small program and must make choices about allocation of resources to achieve some critical mass of work. In all its work, but especially in Prince William Sound, there is a great need for coordination with other research programs (e.g., the *Exxon Valdez* Trustee Council's Gulf Ecosystem Monitoring program and the new North Pacific Research Board).

The Advisory Board has had frank discussions about a number of important issues and problems and has shown a willingness to make changes and institute new procedures when necessary. The committee recognizes that the OSRI Advisory Board is composed of representatives of agencies, each with its own mission and sometimes differing needs. Because of this, the Scientific and Technical Committee (STC) is an important subsidiary body that provides in-depth scientific insight and leadership. The STC is an important part of the checks and balances of the OSRI process. It should have an active role in judging the quality and appropriateness of medium to large proposals. Term limits and clear procedures for selecting new members should be implemented to ensure that the STC remains an independent voice in the OSRI program.

Based on its meeting minutes, the Advisory Board is aware that there is a perception in the science community of conflict of interest and fairness issues within OSRI. This committee was not constituted to investigate these perceptions. However, even if this is no more than an image problem based on outdated or misinformation, the negative perceptions tangibly affect the program's performance. They diminish the program's appeal to qualified outside scientists and cast a shadow over its credibility.

Part of dealing with the negative perceptions about OSRI will necessitate dealing with the close relationship between OSRI and the PWSSC. These two organizations are clearly linked, and this is not necessarily inappropriate. Because OSRI and PWSSC are both small organizations, located in a small and isolated community, there are cost efficiencies to be gained by sharing staff and facilities. However, because OSRI grants significant funding to the PWSSC and because the two organizations share staff, including the director, this is fertile ground to develop perceptions of impropriety. Negative perceptions will take time to overcome, but attitudes can be changed by open communication of the mission and vision and by carefully following all procedures and policies.

Although it is a small program within the larger scientific context and it has some problems that need to be addressed, OSRI is doing good work and it is an important influence in Cordova, Alaska, and the surrounding region. Some of its educational programs deserve special recognition for building strong community partnerships. Regarding the Education and Outreach program, the activities generally are a valuable contribution to the overall OSRI program and should be continued, with emphasis that they have clear links to communicating information about oil spills and their prevention, response, and effects.

There is a fundamental ambiguity about whether research priorities at OSRI flow "top-down" or "bottom up," and this confusion is at least in part why some outside scientists are skeptical of the fairness of the organization's procedures. Is OSRI a fund-granting institution that allows the scientific community to drive the direction and mixture of projects (through the proposal process) under the general guidance of the mission, staff, and Advisory Board? Or is OSRI management responsible for developing a directed science plan that they then implement through directed procurement of specific projects and products? OSRI is now primarily operating in the latter mode. Both approaches have advantages and disadvantages, and neither is inherently better than the other. However, it is likely that many in the science community, as well as other stakeholders, believe that the first approach is what was intended by the authorizing legislation.

If the "top-down" approach is a deliberate choice, OSRI needs to articulate this clearly and honestly, to avoid misunderstandings and disenchantment by proposers. Accordingly, OSRI should be aware that this approach poses the risk of leading to a program that, overall, appeals primarily to the stakeholders at the table and not the broader community in general. It may be that this dilemma is behind the difference of opinion (explored in depth in Chapter 7) about the value of OSRI's modeling efforts: this component is a highlight to many OSRI decision makers, but this committee believes that the current emphasis on real-time spill response capability is inappropriate and perhaps duplicative of other efforts.

These issues should be addressed by the OSRI Advisory Board. It should consider whether it is effective and fair to appear to manage a broad solicitation process when most degrees of freedom are proscribed, and if it prefers the program to be under the strong direction of management, whether there is an adequate mechanism to set priorities and vet ideas.

In summary, OSRI has made considerable progress in its first years, maturing in many ways and resolving some early problems. But this is a critical juncture—a time for evaluation and movement in new directions. The OSRI Advisory Board should play a key role in leading OSRI through a careful strategic planning effort, thus ensuring that the organization has a clear focus in the future.

1

Introduction

In 1989, the T/V *Exxon Valdez* ran aground in Prince William Sound and spilled approximately 11 million gallons of oil. As one of numerous reactions to the spill, Congress passed the Oil Pollution Act of 1990 (OPA 90) and within the legislation mandated the creation of the Oil Spill Recovery Institute (OSRI). OSRI was established to "conduct research and carry out educational and demonstration projects designed to (Title V, Section 5001, Oil Pollution Act of 1990):

- identify and develop the best available techniques, equipment, and materials for dealing with oil spills in the Arctic and sub-arctic marine environment; and
- complement Federal and State damage assessment efforts and determine, document, assess, and understand the long-range effects of Arctic or subarctic oil spills on the natural resources of Prince William Sound and its adjacent waters...and the environment, the economy, and the lifestyle and well-being of the people who are dependent on them, except that the Institute shall not conduct studies or make recommendations on any matter that is not directly related to Arctic or subarctic oil spills or the effects thereof.

OPA 90 identifies the Prince William Sound Science Center (at the time called the Prince William Sound Science and Technology Institute) in Cordova, Alaska, as the administrator of OSRI. Although established in legislation in 1990, substantial funding for OSRI was not provided until 1996. It disbursed its first funds in FY98, at a start-up level of $200,000.

7

Now OSRI receives about $1.2 million per year, generated as interest on a $22.5 million trust held by the U.S. Treasury (U.S. Coast Guard Authorization Act, 1996). Thus, OSRI is a relatively new program with a limited track record to evaluate. To date, the OSRI research program has supported about $5 million of projects over the approximately five years of operation since it began making awards. Figure 1-1 provides a general timeframe of OSRI activities.

Under the guidance of the OSRI director and with advice from its Advisory Board and the Scientific and Technical Committee, the current OSRI R&D grant program translated its legislative mandate into three focus areas: applied technology (to conduct research and development on new technologies for preventing and responding to oil spills in the Arctic and subarctic), predictive ecology (to develop new capabilities to predict changes in populations at risk from spills), and public education and outreach (to make the research process interactive with the public and in general provide public information and education about oil spill impacts and response). These areas of focus and other key aspects of how OSRI is administered arose out of a strategic planning workshop held in 1997, when the Advisory Board provided guidance, adopted procedures, and wrote its first business plan. In this plan, the Advisory Board adopted an allocation system to ensure a balanced program that addressed the OSRI mission, and instructed the director to seek spending targets of 40-40-20 percent for applied technology, predictive ecology, and public education and outreach grants, respectively. These three areas continue to be OSRI's primary areas of focus, although there is some overlap among the areas. The Advisory Board also directed OSRI staff to develop a grant policy manual. OSRI policies are based on the policies and procedures of the National Science Foundation (NSF), the National Oceanic and Atmospheric Administration's (NOAA) National Undersea Research Program, and the *Exxon Valdez* Oil Spill Trustee Council (EVOSTC). The Advisory Board also adopted the current mechanism for grant approval, authorizing the director to approve awards under $25,000, involving the STC in awards of less than $100,000, and requiring Advisory Board approval for awards over $100,000. OSRI funding is authorized for a 10-year period that ends in 2006 and discussions are beginning to determine if the Institute will be continued.

THE COMMITTEE'S CHARGE AND METHODS

The legislators who created OSRI under OPA 90 foresaw the need for periodic reviews to ensure that the program was meeting its legislative mandate, and Section 2731 allowed OSRI to request a review from the National Academy of Sciences. This report is the first external view of

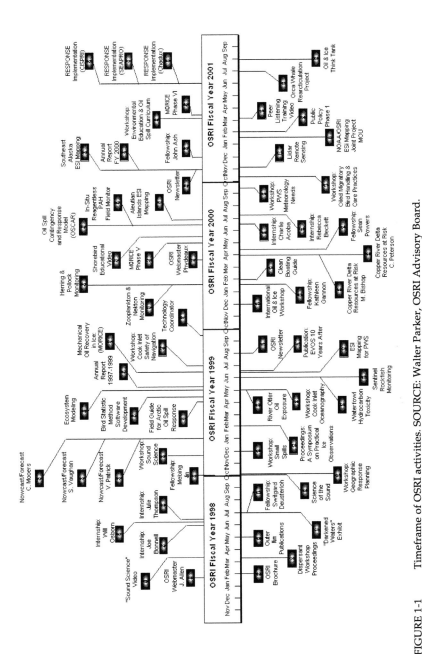

FIGURE 1-1   Timeframe of OSRI activities. SOURCE: Walter Parker, OSRI Advisory Board.

the OSRI program, and it is the effort of a committee of nine members selected based on expertise relevant to the OSRI's program (see Appendix A).

This report is not a project-by-project review of all OSRI activities; it is a broad assessment of the programs strengths and weaknesses, with special emphasis on whether it is addressing its intended mission and whether the work supported is of high quality. In many ways, the committee operated as a visiting committee (an evaluation technique often used in university settings, where a group of outside experts is invited to visit, gather information, and provide an evaluation). The committee met three times over the course of about eight months to gather information, deliberate, and write its report. By necessity, we relied heavily on OSRI to provide documents and answer questions about its modes of operation. As a result, much of our information comes as personal communication and our findings and recommendations are based on our consensus expert judgments.

To gain an understanding of OSRI's mission and activities, the committee held conversations with OSRI staff, the OSRI Advisory Board, members of the OSRI Scientific and Technical Committee, past and current researchers, potential users of OSRI products, and others knowledgeable about OSRI programs. It distributed a call for input to gain other views of the program; although the responses received were anecdotal, they did help shape our understanding of the program. The committee also reviewed selected documents from the OSRI files in an attempt to develop a broad view of how the program has operated and its effectiveness. Documents included but were not limited to:

- the Grant Policy Manual,
- the Advisory Board minutes,
- sample Broad Area Announcements (BAAs),
- samples of proposals received,
- the 1995 Oil Pollution and Technology Plan,
- annual work plans,
- the 1999 Business Plan, and
- progress reports and publications for OSRI projects.

This report is organized to parallel the five questions posed to the committee (Box 1-1). Chapter 1 describes the OSRI mission and provides the general context for this review. Chapters 2 and 3 examine the organization and administration of OSRI and assess whether the process used to select research and technology projects is sound and fair. Chapter 4 considers whether OSRI planning documents set an appropriate course for the future. Chapters 5 through 8 focus on the primary components of the

BOX 1-1
CHARGE TO THE COMMITTEE

The Oil Spill Recovery Institute was established to identify and develop methods to deal with oil spills in the Arctic and subarctic environment and develop a better understanding of the long-range effects of oil spills on the natural resources of Prince William Sound and its adjacent waters, including the environment, economy, and people. This committee was charged to review OSRI's activities (both the research program and technology development and implementation activities). The committee was assigned the following tasks:

• Explore whether the research and activities supported to date adequately address the mission of the Institute,
• Assess whether the research and activities are of good quality and operational aspects are effective and efficient,
• Determine whether the process used to select the research and activities is sound,
• Consider whether existing planning documents set an appropriate course for the future, and
• Offer recommendations for future directions and opportunities suitable to the OSRI mission, and comment, if possible, on mechanisms to increase responses to the organization's calls for proposals.

OSRI program (predictive ecology, applied technology, modeling, and education/outreach) and assesses whether the funded mix of research and technology projects adequately address the mission of OSRI and whether these activities are of high quality. Finally, Chapter 9 summarizes our findings and recommendations.

OSRI STRUCTURE AND FUNCTIONS

OSRI was authorized in OPA 90, but it did not begin to take shape until 1996 when funding was provided within the Coast Guard Authorization Act. This act defined some of the key structures and functions of OSRI. It mandated creating the OSRI Advisory Board and specified its composition, chairmanship, terms, and members' voting status. The operations of the Advisory Board—including policies and procedures related to officers of the board, meetings, quorums, public notice, voting and resolutions, and establishment and operation of an Executive Committee—are detailed in bylaws created by the Advisory Board and amended in April 1999 (Appendix B).

The 1996 legislation also mandated establishment of a Scientific and Technical Committee (STC). This group provides scientific input, and is composed of specialists in matters relating to oil spill containment and clean-up, marine ecology, and the living resources and socio-economics of Prince William Sound and its adjacent waters. The STC provides advice to the Advisory Board regarding the conduct and support of research and technology projects and studies. Appointment procedures for the STC are relatively informal, with people added as the need arises at the request of the director or the Advisory Board and there are no set term limits. The pool of potential members for the STC was identified in the 1996 legislation as "… the University of Alaska, the Institute of Marine Science, PWSSC, and elsewhere in the academic community." Under the bylaws, the chair of the STC is a representative of the University of Alaska and this person is also to serve as a nonvoting member of the Advisory Board.

The legislation mandates that OSRI should operate its granting process using a traditional approach such as those used by the National Science Foundations (NSF) and others. Projects are to be identified by advertising Broad Area Announcements or Requests for Proposals. To evaluate the proposals received, the 1996 legislation requires use of an outside proposal review process "on a nationally competitive basis…." The legislation encourages research results to be published and made widely available, giving the National Oceanic and Atmospheric Administration (NOAA) a role as a repository for copies of all research, educational, and demonstration projects.

The legislation gives responsibility for selecting the OSRI director to the Advisory Board, with recommendations from the STC and the PWSSC. The director is empowered to hire staff and incur expenses. The legislation stipulates various procedures and policies for OSRI's operation and use of funds, and authorized funding for 10 years ending in 2006. Related to funding, it is mandated that the program's funds would come from interest on a $22.5 million endowment. No funds were to be used to institute litigation, purchase real property, or construct buildings, and no more than 20 percent of the funds may be used to lease facilities and administer OSRI. The legislation also stipulated that "the Institute shall not conduct studies or make recommendations on any matter which is not directly related to Arctic or subarctic oil spills or the effects thereof." The revenues for OSRI over its 10-year term are expected to be about $14 million total, with annual revenues averaging $1.4 million.

OSRI AND OTHER GULF OF ALASKA RESEARCH PROGRAMS

OSRI was one of numerous responses to the *Exxon Valdez* oil spill, however, its establishment came at a time when emphasis was shifting away from damage assessments and toward improving our capabilities to understand and respond to future events. Several external influences were important in how the OSRI legislation was implemented as an operational program. For example, according to conversations with Advisory Board members involved in the founding of OSRI (W. Parker, OSRI Advisory Board, personal communication, February 7, 2002), a major research program known as GLOBEC (Global Ocean Ecosystem Dynamics) was evolving at the time OSRI was being established in the early 1990s and the GLOBEC experience was considered in designing OSRI. GLOBEC stressed the importance of understanding the coupling of physical and biological systems in understanding ecosystem change in marine systems. Also in the same timeframe, the *Exxon Valdez* Oil Spill Trustee Council (EVOSTC) was administering a major program designed to conduct damage assessments after the spill and conduct research related to restoration of damaged resources and understanding of environmental change in the northern Gulf of Alaska. One part of the EVOSTC activities was a program called the Sound Ecosystem Assessment (SEA)[1] and a 1997 SEA workshop and a 1997 SEA workshop on modeling appears to have been important to OSRI planning.

With about $1 million in allocatable funding each year, OSRI is a relatively small program. But it is working in a region with many other ongoing research programs (Box 1-2) and coordination and leveraging of funding is an essential part of making sure that the various programs are synergistic and not redundant.

---

[1]The Sound Ecosystem Assessment program was a multidisciplinary, ecosystem-level investigation of factors affecting recovery of pink salmon and Pacific herring in Prince William Sound, Alaska, following the 1989 oil spill. The primary avenues of investigation were descriptive oceanography and ocean circulation modeling, to investigate the roles of physical forcing on production of food, and ecosystem modeling, to develop analytical and predictive capabilities for assessing likely effects of perturbations to the ecosystem. SEA involved 13 coordinated research projects led by investigators from the University of Alaska Fairbanks Institute of Marine Science, the Prince William Sound Science Center, the Alaska Department of Fish and Game, the U.S. Forest Service Copper River Delta Institute, and the Prince William Sound Aquaculture Corporation. It was sponsored by the EVOSTC.

BOX 1-2
EXAMPLES OF RELEVANT RESEARCH ACTIVITIES IN THE
PRINCE WILLIAM SOUND AND GULF OF ALASKA REGION

• **U.S. Global Ocean Ecosystems Dynamics Program (GLOBEC) (funded by NSF and NOAA).** The U.S. GLOBEC research initiative has been called for by the oceanographic, marine ecology and fisheries communities to address the question: what will be the impact of changes in our global environment on populations and communities of marine animals comprising marine ecosystems? The U.S. GLOBEC approach is to develop basic information about the mechanisms that determine the variability of marine animal populations. Through such understanding scientists can produce reliable predictions of population changes in the face of a shifting global environment. More information is available online at <http://www.nsf.gov/geo/egch/gc_globec.html>.

• **Ocean Carrying Capacity (NMFS).** ABL's Ocean Carrying Capacity program continues the NMFS role in the stewardship of living marine resources of the North Pacific Ocean. This research, supported by the North Pacific Anadromous Fish Commission, will bridge the gap between ongoing coastal ecosystem studies in Prince William Sound and the high seas Carrying Capacity and Climate Change study developed by North Pacific Marine Science Organization (PICES). More information is available online at <http://www.afsc.noaa.gov/abl/OCC/occ.htm>.

• **Steller Sea Lion Coordinated Research Program (NOAA and others).** The listing of the Steller sea lion (*Eumetopias jubatus*) as a threatened species under the Endangered Species Act in 1990 created new challenges for fisheries managers in the National Marine Fisheries Service and the North Pacific Fisheries Management Council. Managers must balance between two sometimes conflicting objectives: protecting and aiding the recovery of the Steller sea lion under the Endangered Species Act while at the same time providing for sustainable and economically viable fisheries under the Magnuson-Stevens Fishery Conservation and Management Act. The Steller Sea Lion Coordinated Research Program, administered by the National Marine Fisheries Service (NMFS) Alaska Fisheries Science Center, is composed of more than 150 research projects conducted on Steller sea lions. To date, roughly $84,000 million has been appropriated for this program to coordinating agencies such as NOAA (NMFS, NOS, and OAR), the University of Alaska, North Pacific Universities Marine Mammal Research Consortium, Alaska Department of Fish and Game, Alaska SeaLife Center, and the North Pacific Fisheries Management Council. More information is available online at <http://www.afsc.noaa.gov/Stellers/coordinatedresearch.htm>. (NOAA's webpage is currently under construction.)

- **EVOSTC GEM Program and Transition Projects.** GEM is a long-term commitment to understanding the Gulf of Alaska and sharing information that will determine the future of the gulf ecosystem and the human activities that depend on it. What makes GEM unique is that it incorporates inter-agency cooperation and collaboration, public involvement and accessibility, and informative data and information on the Gulf of Alaska ecosystem. More information is available online at <http://www.oilspill.state.ak.us/gem/index.html.>

- **NASA Sea-viewing Wide Field-of-view Sensor (SeaWiFS) (MODIS).** The purpose of the SeaWiFS project is to provide quantitative data on glo-bal ocean bio-optical properties to the Earth science community. The SeaWiFS project has been designated to develop and operate a research data system that will gather, process, archive, and distribute data received from an ocean color sensor. This program evaluates the dynamic nature of the world's oceans and climate, and the importance of the ocean's role in global change. More information is available online at http://seawifs.gsfc.nasa.gov/SEAWIFS/BACKGROUND/.

- **Prince William Sound Regional Citizens Advisory Council (PWS RCAC).** The PWS RCAC is an independent, non-profit organization formed in 1989 after the *Exxon Valdez* oil spill. The council provides a voice for communities and citizens on oil industry decisions that may affect them. The council's members are communities and groups affected by the *Exxon Valdez* oil spill. The council is certified under the federal Oil Pollution Act of 1990 as the citizens' advisory group for Prince William Sound. The council conducts independent research to support its mission and goals. One such example is its Long-Term Environmental Monitoring Program (LTEMP), a modified mussel-watch program that monitors for the presence of hydrocarbons resulting from Alyeska Marine Terminal and associated tanker operations. More information is available online at <http://www.pwsrcac.org>. A similar Citizens' Advisory Council exists for Cook Inlet.

- **Coastal Ocean Processes (CoOP) program (NSF, NOAA, ONR).** The coastal ocean has a number of unique physical and meteorological pro-cesses that promote high biological productivity, active sedimentary pro-cesses, dynamic chemical transformations, and intense air-sea interactions. As more of the world's population shifts towards coastal areas, human impacts on the coastal ocean in terms of pollution, waste disposal, and recreation continue to increase. CoOP research provides a greater under-standing of how the coastal ocean system functions. Research projects con-ducted under the auspices of CoOP are funded by the National Science Foundation, the Office of Naval Research and the National Oceanic and

BOX 1-2 Continued

Atmospheric Administration.   More information is available online at
<http://www.geo.nsf.gov/cgi-bin/showprog.pl?id=49&div=oce>.

• **North Pacific Research Board (NPRB).** The NPRB's mission is to
develop a comprehensive science program of the highest caliber that provides better understanding of the North Pacific, Bering Sea, and Arctic
Ocean ecosystems and their fisheries. The board's research priorities include fish habitat, ecosystems dynamics, endangered and stressed species,
reduction in fishing capacity and bycatch, salmon dynamics, fish stock
assessment and ecology, and contaminants. Its work will be conducted
through science planning, prioritization of pressing fishery management
and ecosystem information needs, coordination and cooperation among
research programs, competitive selection of research projects, enhanced
information availability, and public involvement.

2

Organization and Administration

Implementation of the Oil Spill Recovery Institute's (OSRI) mission and accomplishment of the objectives of the authorizing legislation depend on having an effective organizational structure, and in particular having unbiased administrative procedures for the award of grants and contracts. In recognition of the finite nature of OSRI resources, it's important that OSRI interact effectively with other research programs in the region to avoid duplication of efforts and provide synergy in the accomplishment of common objectives. OSRI products must be effectively communicated to users and stakeholders. To this end, OSRI must be an objective and fair partner to a variety of people.

To understand the process that has been developed to manage and distribute OSRI funding, the various steps in the process and the role of oversight and advisory bodies must first be understood. The authorizing legislation sets forth an organizational structure that includes a director, an Advisory Board, and a Scientific and Technical Committee (STC). The Advisory Board determines the policies for the conduct and support of research, projects, and studies through contracts and grants "awarded on a nationally competitive basis." The OSRI Advisory Board directed that awards be competed on a national level, meaning that they be advertised nationally and competed by the largest possible pool of potential researchers, not just researchers from the immediate area. In general, national competitions are ways of encouraging submission of high-quality proposals.

With guidance from its advisors, OSRI has developed a fairly typical process to gather and review proposals and make awards. Key elements of this process include the development of Broad Area Announcements (BAAs), development of annual work plans, and peer review of proposals. The grant process is described in more detail in Chapter 3. This chapter examines the oversight structure more broadly, looking at roles, fairness, and effectiveness and whether adequate checks and balances are in place to ensure the integrity of the process and produce an atmosphere where partnerships and collaborations can flourish.

THE ADVISORY BOARD

The main oversight committee of OSRI is its Advisory Board, which has a key role in setting policy. The Advisory Board is composed of 16 members as defined in the authorizing legislation (see Table 2-1 or Appendix E). The Advisory Board is chaired by the representative of the U.S. Secretary of Commerce. Board members include state and federal employees. Several board members are nominated by interested parties through the governor of the State of Alaska. At large representatives are considered and named by the Advisory Board itself. In addition, there are two nonvoting representatives from the Institute for Marine Sciences and the Prince William Sound Science Center (the director of PWSSC, who also serves as director of OSRI, serves as a nonvoting representative).

One important role of the Advisory Board, carried out by a subcommittee in consultation with staff, is preparation of annual work plans. These are important documents because they articulate the goals for the program.

In examining the roles and responsibilities of the various oversight groups, we conveyed many questions to OSRI to be sure we had a clear understanding of how they operate. The most relevant of these discussions are included here as a summary of how OSRI operates.

- **Do members of the Advisory Board serve on other bodies or committees of the OSRI or the PWSSC?**

OSRI has an Executive Committee, Finance Committee, and Annual Work Plan Committee (see Appendix E). Members are elected to the Executive Committee, per the bylaws, and appointed to the other committees, generally on an annual basis. One voting member of the OSRI Advisory Board, Gail Evanoff, also currently serves on the PWSSC Board of Directors (see Appendix E). She was appointed to the OSRI Advisory Board as one of the Alaska Native representatives by Governor Tony Knowles. There is one nonvoting member of the PWSSC Board of Directors on the OSRI Advisory Board, per OPA 90 legislation. Otherwise, there is no overlap between the OSRI and PWSSC boards.

- **How is the working plan committee selected?**

The work plan committee is selected by the Advisory Board, usually at its winter meeting with a report/recommendation being presented in the late summer/early fall meeting for the next fiscal year's plan. Appointment is for one year. Declarations of conflict of interest are done by all Advisory Board members when they first join the board (see Appendix B, OSRI bylaws, for the form that all board members complete). *Specific conflicts that may arise during annual work plan discussions are declared at that time.*

SCIENTIFIC AND TECHNICAL COMMITTEE

To provide more detailed scientific input, the Advisory Board empanels a Scientific and Technical Committee (STC) composed of specialists in matters related to oil spill containment and cleanup technologies, Arctic and subarctic marine ecology, and the living resources and socioeconomics of Prince William Sound and its adjacent waters. This committee is to be composed of scientists from the University of Alaska, the Institute of Marine Sciences (IMS), the Prince William Sound Science Center, and elsewhere in the academic community (see Table 2-1 or Appendix E). The STC provides advice to the Advisory Board as requested and it makes recommendations regarding the conduct and support of research and technology projects and studies. The process for the selection of STC members, the terms of service, and the schedule for rotation is unspecified. The committee's questions and OSRI staff responses about the STC included the following:

- **Are members of the Scientific and Technical Committee a subset of the Advisory Board?**

With the exception of the STC chair, who is the academic appointee to an ex officio Advisory Board position, all STC members are outside experts from academia, management, industry, and the public, per OPA 90 requirements (the legislation states these specialists must come from "the University of Alaska, the Institute of Marine Science, the Prince William Sound Science Center, and elsewhere in the academic community."

- **Do STC members serve on other committees or bodies of OSRI or the PWSSC?**

No STC members serve on other OSRI or PWSSC committees or bodies, other than the STC chair, who is a nonvoting member (currently John Goering, who is the IMS/UAF representative on the Advisory Board).

TABLE 2-1   Membership on Oil Spill Recovery Institute and Prince
William Sound Science Center Advisory Groups as of Spring 2002

| NAMES (as of June 2002) | OSRI | | | |
| | Advisory Board | | | |
	Board Member	Executive Committee	Finance Committee	FY03 Work Plan Committee
Virginia Adams	✓			
John Allen				
Ed Backus				
Chris Blackburn				
John Calder	✓	✓ [a]		
Capt. Jack Davin	✓			
Douglas Eggers				
Gail Evanoff	✓			
Mark Fink	✓			✓
Carol Fries	✓		✓	
John Goering	✓			
John Harville				
Raymond Jakubczak				
Meera Kohler				
R.J. Kopchak	✓	✓	✓ [a]	
Marilyn Leland	✓	✓		✓
Calvin Lensink				
Doug Lentsch	✓			✓

Scientific and Technical Committee	PWSSC		
	Board of Directors		
	Executive Committee	Board Member	Emeritus Members
			✓
			✓
		✓ [b]	
✓			
		✓	
✓ [a]			
			✓
✓			
	✓		
			✓

TABLE 2-1 Continued

	OSRI			
	Advisory Board			
NAMES (as of June 2002)	Board Member	Executive Committee	Finance Committee	FY03 Work Plan Committee
Simon Lisiecki				
Ole Mathisen				
Trevor McCabe				
Charles Meacham				
Alan J. Mearns				
Doug Mutter	✓	✓[b]		
Stu Nozette				
Charles Parker				
Walter Parker	✓			
Leslie Pearson	✓	✓		✓
Stanley (Jeep) Rice				
Thomas C. Royer				
Steven Taylor				
Gary Thomas	✓[c]	✓[c]		
Ed Thompson	✓	✓	✓	✓
Mead Treadwell			✓[c]	
Glenn Ujioka	✓			

	PWSSC		
Scientific and Technical Committee	Board of Directors		
	Executive Committee	Board Member	Emeritus Members
		✓	
		✓	
		✓	
	✓		
✓			
		✓	
	✓ [b]		
	✓ [a]		
✓			
✓			
		✓	
✓ [c]		✓	
	✓		

TABLE 2-1   Continued

| NAMES (as of June 2002) | OSRI | | | |
| | Advisory Board | | | |
	Board Member	Executive Committee	Finance Committee	FY03 Work Plan Committee
Nolan Watson				
David Witherell				
Edward Zeine				

^{a} chair.
^{b} vice chair.
^{c} nonvoting member.

- **Are there term limits for STC membership and who decides this?**

Duration has been until they resign or another recommendation is made to the Advisory Board for membership on the STC. Appointment is officially by the OSRI director with concurrence/approval by the Advisory Board as a whole. We have not asked STC members to fill out a conflict of interest form, but the first paragraph of all proposal review forms requests such conflicts to be declared and for the reviewer to return the proposal unreviewed if there is a conflict.

- **Can you provide a list of the OSRI management team and the PWSSC management team?**

OPA 90 established PWSSC as the administrative institution for OSRI. All administrative positions for OSRI are PWSSC employees. Owing to the small size of the staff and budget and the need for several specialties to operate OSRI and PWSSC programs, the administrative positions for OSRI and PWSSC are shared: G.L. Thomas, director/president, Nancy Bird, administrator/vice president, Penny Oswalt, finance director, and Jessica Miner, bookkeeper. All administrative positions are full-time but the salaries are shared 2/3 to 1/3 between OSRI and the PWSSC's general fund, with one exception, director/president G.L. Thomas, who is also supported on non-OSRI research grants and contracts (the OSRI administrative staff is not allowed to apply for OSRI grants). G.L. Thomas's position has broken down to about 1/3 PWSSC administration, 1/3

	PWSSC		
Scientific and Technical Committee	Board of Directors		
	Executive Committee	Board Member	Emeritus Members
			✓
	✓		
	✓		

NOTES: This table shows membership in the full suite of oversight bodies, and uses names current as of spring 2002 as a way to compare membership of the groups to determine if there is an over-reliance on the same advisors. There appears to be good diversity of participation although the OSRI director has many roles. Appendix E provides membership lists for each body separately.

OSRI administration, and 1/3 non-OSRI research grants over the history of the institute. Part-time administrative positions for network administration, administrative assistant, librarian and receptionist are arranged on need, budget, and availability basis.

DEVELOPMENT OF BROAD AREA ANNOUNCEMENTS

Solicitation of proposals for OSRI projects is done mostly through the issuance of BAAs. An important step in the procurement process is the development and writing of these BAAs, because they are the instructions by which potential proposers understand the intended scope and focus of desired research and write their applications for funds. BAAs are written by OSRI staff. Chapter 3 looks in detail at grant award policies and suggests improvements. This section answers some key questions that arose as the committee sought to understand the administration hierarchy.

The following questions/issues about BAAs were identified by the committee. Responses from OSRI staff follow the questions.

- **Are BAAs approved/reviewed by others, such as the Advisory Board or the STC?**

Before February 2002, BAAs were written by staff and reviewed/approved by the OSRI director. After February 2002, the STC must review/approve BAAs before release.

• **Are authors and/or reviewers of BAAs eligible to submit proposals in response to those BAAs?**
No.

• **Why are BAAs used when Requests for Proposals (RFPs) might be more appropriate, and is the use of BAAs consistent; when are open BAAs chosen and why?**
OSRI has used both BAAs and RFPs in the grant process. Generally, when a specific task/function/effort is identified, an RFP is used. BAAs are used for projects where a broader scope of proposals is desired. Open BAAs are used in the education program where OSRI seeks to identify candidates for fellowships/internships and to fund science workshops as part of our continuing planning program.

REQUIREMENTS FOR PROPOSALS

Beyond the requirements for being responsive to the substance of the BAA or RFP, additional requirements are sometimes stipulated by OSRI. There are two things that may discourage potential applicants: a cap on allowable overhead and a call for matching funds that many people interpret as a requirement. Specifically, OSRI policy limits overhead to 20 percent and it requests a two-for-one match, although the match may be in-kind. The committee asked OSRI staff the following questions about proposal requirements:

• **Who decided and why was it decided to cap overhead at 20 percent?**
A 20 percent overhead cap for the administration of OSRI was established by OPA 90. The OSRI Advisory Board adopted this as a guideline also to be used for indirect costs for outside proposals after lengthy debate (see the minutes of August 12, 1998 - pages 6 and 9). However, the board allows the director discretion to approve higher than 20 percent indirect costs based on the project's merit. As a practice, the director approves requests from academic institutions for higher rates when the institution has a government-audited indirect cost rate.

• **Is any preference given to proposals that have a Prince William Sound Science Center (PWSSC) co-principal investigator (PI)?**
No. The Advisory Board specifically reviewed such clauses, which are common at other funding programs in Alaska, such as the UAF-CMI program (fed-

eral funds), the UAF-PCC program (industry funds), and others, and they decided that this should not be done. In fact, the Advisory Board adopted a policy with the opposite intent: when proposals are equally ranked, a proposal with a PWSSC PI is funded secondarily to a proposal without a PWSSC PI. Any PWSSC proposal that is funded by OSRI must be clearly superior to proposals from non-PWSSC PIs

The 20 percent cap on overhead, even though it can be waived by action of the OSRI Advisory Board or director, is a real limitation to the program. By current standards, this is a low allowable rate and seeing this on the website or in a BAA or RFP would discourage scientists from many institutions from applying for grants. Thus, the cap is acting to reduce the diversity of proposals received, and the lack of wide competition is a detriment to the program. The objective of seeking matching funds, to leverage OSRI's impact, is sound and can give the program broader impact, but care should be taken to communicate the fact that this is a goal but not a requirement.

UNSOLICITED PROPOSALS

Although OSRI staff discourages unsolicited proposals, they are accepted if received. If an unsolicited proposal is judged to warrant consideration, it is translated into a BAA and advertised. The original proposer receives no favorable treatment in the process. The committee asked OSRI staff the following questions/issues about unsolicited proposals:

- **Who decided the policy for unsolicited proposals?**
The OSRI Advisory Board decided policy for handling unsolicited proposals.

- **Who decides if an unsolicited proposal is translated into a BAA?**
The Advisory Board decides. The staff discourages unsolicited proposals. We encourage public input at Advisory Board meetings for project ideas to be considered annually by the Work Plan Committee.

NATIONAL COMPETITIVENESS

The legislative mandate is for nationally competitive award of OSRI resources. The following question was asked of OSRI staff to clarify advertising procedures.

- **What procedures are used to ensure nationally competitive awards?**
Funding opportunities are advertised through a variety of media to attract proposals from a broad pool of applicants. The Commerce Business Daily *and*

the Internet are the primary methods for advertising solicitations. Newspapers, direct mailings, e-mail lists, and postings of opportunity at professional meetings and e-mail networks, such as LABNET, are used where deemed appropriate by OSRI staff.

CONFLICT OF INTEREST AND FAIRNESS

During this review, the committee's members and some of those who responded to our call for input raised questions about how conflict of interest and fairness issues were managed in the OSRI award process. For example, some members of the relevant science communities perceive that personnel involved in the promulgation of BAAs, the peer review process, and the selection process can also have proposals under consideration by the same process. However, the OSRI Grant Policy Manual has sufficient rules to prohibit this problem, leaving the committee to wonder whether this perception is outdated or misguided.

There have been problems with unnecessarily limited circulation of BAAs and inappropriately short deadlines (see Chapter 3), which some might consider a fairness issue but which the committee sees as administrative challenges that can be corrected. In addition, as mentioned earlier, restrictions on overhead charges and the perception that matching funds are required, rather than just encouraged, also acts to limit respondents. Confusion over whether those serving on the OSRI Advisory Board and the Scientific and Technical Committee can receive OSRI funds also causes negative perceptions about the fairness of the process.

The OSRI authorizing legislation was quite prescriptive in setting up some of the organization's rules and procedures and creates a situation that may look flawed unless everyone involved is extremely careful to avoid both real and perceived conflicts of interest. For example, if it had not been required by legislation, it is unlikely that a grant-awarding organization like OSRI would be administered jointly with a grant-receiving organization like the Prince William Sound Science Center.

The director of OSRI is in a particularly sensitive position in terms of being open to perceptions of unfairness in the process or conflict of interest. Although the OSRI director is precluded from being a principal investigator on OSRI-supported projects, the perception problem arises because the OSRI director is also the director of the PWSSC and serves as a principle investigator or co-PI on PWSSC projects.

The National Science Foundation (NSF), the Office of Naval Research (ONR), and others have considered ways to allow their program managers to remain active scientists but with clear boundaries that protect their programs from conflict of interest issues (i.e., the Independent Research

Program at NSF and the Research Opportunities for Program Officers program at ONR), and OSRI might develop ideas from those mechanisms.

Procedures also dictate that whoever is the director of PWSSC will be designated as a representative on several boards and committees. Selection of the membership on various committees and the OSRI Advisory Board appear to be strongly influenced by the director's recommendations, which again contributes to questions regarding fairness since these bodies are intended to provide independent oversight. The director, as would be expected, materially participates in and approves the content of BAAs and allocation of awards after review and will always need to take extreme care in representing each organization and managing conflicting duties. By design the same person leads both a center that conducts research and a research grant-awarding program. The committee can only caution that whoever serves in the role must be aware of the image issue and be proactive in combating problems by always following fair and open procedures and ensuring that all actions are transparent to other organizations and colleagues throughout the science community.

As illustrated in Table 2-1, there is not significant overlap in use of people in different advisory roles, except for the expected overlap of OSRI Advisory Board members serving on its Executive Committee, Finance Committee, and Work Plan Committee and except for the many roles assigned to the OSRI director. Good variety among the advisors is important because using the same people tends to limit creativity in any program. Another constraint on creativity is lack of organized rotation of participants, to allow room for new ideas and perspectives. As possible, given the legislative requirements for some of the roles, scientists with broad perspectives should be sought (not just scientists with specialized sub-Arctic science experience).

3

Grant Award Policies and Procedures

The Oil Spill Recovery Institute (OSRI) provides its funds to outside investigators primarily through an open, competitive process (with some exceptions, as discussed in Chapter 6). OSRI's grant policies are outlined in its Oil Spill Recovery Institute Grant Policy Manual. The current edition is dated January 2002 and reflects revisions as the OSRI program has evolved. The Oil Spill Recovery Institute modeled its basic grant policies and procedures on those used by the National Science Foundation (NSF) and the National Oceanic and Atmospheric Administration's (NOAA) Undersea Research Program. Proposals are solicited primarily through Broad Area Announcements (BAAs), although Requests for Proposals (RFPs) are sometimes used.[1] The content of the BAAs varies each year based on recommendations by the OSRI Advisory Board. According to the Grant Policy Manual, "cooperative" proposals between researchers

---

[1] As applied by most agencies and research programs, Broad Area Announcements (BAAs) set out a broad goal and rely on those who propose activities to define how they will achieve the goal. For instance, this is how DOD and NSF move forward on questions of infrastructure or equipment procurements. The proposer tells what is needed, how it addresses the goal, and why it is important, and the program administrators judge which proposal best suits their needs. In contrast, Requests for Proposals (RFPs) are generally more proscriptive, outlining goals and objectives more explicitly, but generally also rely on the proposer to develop the plan. An example is the NSF call for proposals for its Biocomplexity Initiative. Finally, contracts are used when specific services and defined deliverables and products are needed, such as computer systems, publication services, and routine data collection. OSRI has at times blurred the lines between these types.

and users are encouraged with cost sharing being used as one metric of user interest. In addition, multidisciplinary and multi-institutional projects are encouraged. Funding may be in the form of grants, contracts, or cooperative agreements. Awards vary by size and level of approval required: the January 2002 edition of the Grant Policy Manual gives the director authority to approve small awards (less than $25,000 per year), but additional approval by the Scientific and Technical Committee (STC) is now required for medium-size awards ($25,000 up to $100,000 per year), and by the OSRI Advisory Board for large awards (e.g., $100,000 per year). Additional general information on policies can be found in the Grant Policy Manual (OSRI, 1998, 2002b).

PROCESS TO SOLICIT PROPOSALS

BAAs and Other Announcements

According to the OSRI Grant Policy Manual, BAAs are to be "issued annually by the OSRI to solicit proposals" within the three program areas. The manual does not provide specific timelines for BAAs or other policies related to the circulation of BAAs. BAAs are generally advertised in the *Commerce Business Daily*, various newspapers, local television, radio, and sometimes announcements at regional, national, and international conferences or through e-mail networks such as LABNET, EVOS Trustee Council mail lists, and more. Three sample BAAs are included in Boxes 3-1, 3-2, and 3-3, and additional examples are included in Appendix F.

The committee has concerns about how the solicitation process is used. First, there is no pattern or regularity to these announcements, and the amount of time allowed for response varies from announcement to announcement. This lack of predictability or regularity in advertisements and deadlines diminishes the likelihood that high-quality researchers not currently affiliated with or funded by OSRI will see opportunities to participate. A nationally competitive program will be achieved only by encouraging broad participation from leading figures. OSRI needs to implement improvements in the writing and advertising of its solicitations for proposals. All solicitations should be open for more reasonable amounts of time (e.g., three months) following broad advertising to allow for maximal response by potential proposers.

Solicitations for research proposals (whether BAAs or RFPs) have not always been publicly announced long enough for adequate responses to be generated. One example is the BAA seeking proposals related to Biological Monitoring of Herring and Pollock in Prince William Sound: it was posted on December 15, 1999 and had a closing deadline of January 7, 2000 (Box 3-1). In addition to the brief opening, this solicitation included

BOX 3-1
Broad Area Announcement for
OSRI's Predictive Ecology Program

Deadline for receipt of applications: January 7, 2000.
Date this BAA was posted: December 15, 1999.

Request for Proposals—Biological Monitoring (Herring and Pollock in PWS)

The Oil Spill Recovery Institute (OSRI) is seeking proposals to conduct echointegration surveys of herring and pollock biomass in Prince William Sound (PWS) during February-March of 2000. OSRI anticipates staging two surveys out of Cordova, Alaska, with each survey lasting one to two weeks in duration. Precise dates will depend upon weather and fish behavior. The maximum amount of funding for this contract is $60,000.

Professional services sought through this RFP include the survey pre-logistics (design, equipment preparation, etc.), data acquisition, data analysis, and reporting of acoustic results. Data acquisition and analysis will require coordination with Alaska Department of Fish and Game (ADF&G). ADF&G personnel are responsible for vessel charter and biological sampling of the acoustic targets. The selected contractor will provide, install, and operate all scientific echosounding necessary for estimation of pollock and herring biomass equipment onboard a vessel provided by ADF&G. Scientific acoustic equipment should consist of: 1) a current model echosounder, operating at 38-120 kHz simultaneously with a GPS receiver for the collection of time-linked, geographic-coded acoustic data, 2) a transducer mounted in a downward-looking configuration within a towed body, 3) all necessary computer equipment and software to acquire and process data. Additionally, cooperation with other researchers and technologists that OSRI might employ is deemed essential.

After the completion of field data collection the selected contractor will work with ADF&G to identify and convert acoustic data into fish density and biomass (length frequency, length-weight relationships, species composition, etc.). The contractor will prepare a preliminary and final report detailing the acoustic survey methods, the analysis of acoustic signals and the estimated biomass present in the survey area. An electronic copy of the 2d arrays of acoustic data from survey transects in ASCII file format is also a requirement. The final report and data file will be due on the July 30 following the survey.

As part of the collaborative agreement between OSRI and ADF&G, the selected contractor is required to make copies of the final report and data set available to ADF&G on July 30th. Preliminary reporting to ADF&G will be made at the earliest possible convenience after the survey. This information will be used by ADF&G for fisheries management.

The design will use all available aerial, vessel, and historical information on the distribution of fish to reduce size of the survey to a reasonable area. Where fish concentrations are found, repeated nighttime surveys will be conducted over their concentrations until enough "good" repetitions are made to reduce the influence of truncation. On each survey as many near-parallel acoustic transects should be made as possible in the area that covers a known concentration of fish. After the fish are repeatedly surveyed, a coordinated effort will be made with ADF&G to target known concentrations of the fish monitored for biological information. Every large concentration of fish must be sampled effectively with nets to insure accuracy of the survey information.

Prospective contractors interested in submitting proposals should direct technical questions and project proposals to the OSRI staff (Gary Thomas, Walter Cox or Nancy Bird) at P.O. Box 705, Cordova AK 99574, (907) 424-5800. Proposals should contain the name and address of the firm; name, address, phone and fax numbers of the contact person for the proposal; a comprehensive description of the equipment and procedures to be used; as well as a brief description of the qualification of the firm and key personnel. Experience of key personnel is critical. The deadline for proposals is January 7, 2000, and proposals must be received by the deadline to be considered. Contractors wishing to submit a proposal are advised that qualification and capabilities will be considered in the evaluation process. Specific criteria that will be used to evaluate proposals include:

1. The capabilities, related experience, facilities, equipment, techniques and methods of the proposing organization.

2. The capabilities, qualifications and experience of the proposing organization's key personnel involved in actively performing the data acquisition, analysis and reporting.

3. The organization's record of past performance with similar types of projects. Applicants should include references to aid in evaluation of performance criteria.

4. The organization's estimated cost to perform the required professional services.

5. Proposers that anticipate problems with meeting the following timeline for the project, or that are inflexible with the delays that might be caused by inclement weather should address these concerns in their proposal.

Pollock survey schedule:
December 15, 1999—Solicitations issued
January 7, 2000—Final day for submission of proposals
January 14, 2000—Award contract
January 2000—Presurvey logistics (calibration and field trials)

BOX 3-1 Continued

February 17, 2000—Install equipment
February 20, 2000—Begin acoustic survey (approximately two weeks)
April 1, 2000—All biological data available from ADF&G
July 2000—Draft final report and data file due
December 2000—Final report as a manuscript submitted for publication

Herring survey schedule:
December 15, 1999—Solicitations issued
January 7, 2000—Final day for submission of proposals
January 14, 2000—Award contract
January 2000—Presurvey logistics (calibration and field trials)
March 17, 2000—Install equipment
March 20, 2000—Begin acoustic survey (approximately two weeks)
June 1, 2000—All biological data available from ADF&G
July 2000—Draft final report and data file due

an extremely detailed schedule and was awarded to a former colleague of the director, which added to perception problems in the science community. One announcement seeking a workshop facilitator was posted on March 29, 2002, with an original deadline of April 19, 2002 (although this was subsequently extended to June 3, 2002), showing that the practice was still occurring as of June 2002.

OSRI generally discourages unsolicited proposals, but the Grant Policy Manual outlines procedures for dealing with them. Unsolicited proposals falling "within the scope of OSRI's mission" may be evaluated through the normal review process. Unsolicited proposals that are determined to fall outside the immediate scope of open BAAs, but are of interest to OSRI, may lead to the issuance of a related BAA. The unsolicited proposal would then be reviewed along with any others generated through the BAA process. This policy may discourage novel science or technologies because any novel areas identified by potential researchers would then be made public through the formal review process that followed. Some unique opportunities may be lost by not having a mechanism to deal with unsolicited proposals. One approach would be to have a periodic open BAA (e.g., once every two years) that specifically solicits novel studies related to the OSRI mission.

For solicited proposals, the nature of the solicitations is intended to be steered by the annual work plans, with approval from the Advisory Board. In its review, the committee was surprised by the great variability in the depth of information provided in the BAAs examined. Box 3-2 is a BAA that solicits proposals for modeling of mesoscale atmospheric conditions in Prince William Sound, and the text describing what is clearly a complex need is extraordinarily brief. The BAA for atmospheric modeling of Prince William Sound called for a specific application for a particular need OSRI had identified. Thus, it would be more appropriately solicited with an RFP stating the requirements. Also, the total text of the second paragraph is quite short. In the first example (see Box 3-1) the text describing what was needed for biological monitoring of herring and pollock in Prince William Sound is extensive and detailed. In fact, the committee was surprised that the biological monitoring solicitation was considered a BAA, given its proscriptive nature. A BAA for such a project, if designed to bring forward a variety of thinking, might have been phrased to request: "Provide a survey of important biomass components of Prince William Sound in a manner compatible with their use in OSRI-funded modeling efforts." The review process would then select the best ideas.

BOX 3-2
Broad Area Announcement for OSRI's Applied Technology

Deadline for receipt of proposals: July 15, 2001.
Date this BAA was posted: not available.

Atmospheric Modeling of Prince William Sound
    The Oil Spill Recovery Institute (OSRI) is soliciting proposals for modeling of mesoscale atmospheric conditions within Prince William Sound (PWS). This primary objectives sought through this solicitation are the design, development and validation of an atmospheric model (or selection, implementation and validation of a suitable existing model) for an environment with significant orographic effects. Additional objectives include: design of required initialization data sets; development of a graphical user interface; design of a data distribution scheme; and collaboration with other OSRI funded researchers in meeting OSRI objectives. The selected model will interact with other existing OSRI numerical models describing ocean currents (POM), oil spill dispersion and subsurface plume trajectories.
    The work conducted through this advertisement will contribute to the development of a Nowcast/Forecast system for PWS incorporating physical and biological ecosystem dynamics within a three dimensional real-time ensemble of numerical models. A total of $100,000 dollars per year is available through this solicitation. The awarded contract will be renewable on an annual basis. Proposals providing for a supported FTE within a university, state or federal organization or agency will be considered.

BOX 3-3
Broad Area Announcement for OSRI's Applied Technology
Program

Deadline for receipt of applications: December 31, 1999.
Date this BAA was posted: October 6, 1999.

The OSRI announces a competition for *Computer Simulation of the Spatial-Temporal Distribution and Impacts of Dispersed and Non-Dispersed Oil Spills*

The Oil Spill Recovery Institute is accepting proposals for technologies that demonstrate three dimensional trajectories as well as the resulting physical and biological environmental impacts of dispersed and non-dispersed oil in arctic and subarctic marine environments. Proposals that focus on Alaskan oil transportation routes and utilize realistic models will be given preference.

The total FY99 budget for this program area is $200,000 and OSRI anticipates funding one or more projects from this total amount. The duration of the grant awards will be for one year with an option to renew.

Box 3-3 provides another example. This BAA is extremely brief and vague, particularly for a program with a $200,000 budget. There are only two sentences and a title indicating the subject area to potential researchers. The title and text imply that the BAA is requesting simulations of the spatial-temporal distribution and impacts of dispersed and nondispersed oil spills. The BAA also states that both physical and biological environmental impact modeling are to be demonstrated as a capability. This suggests that the models should have already been developed and that a simulation study is requested.

There is no indication in the Box 3-3 BAA that the funding was intended as a purchase of an oil spill model for OSRI that might be used to run simulations in PWS as part of the Nowcast/Forecast system, which is ultimately what was done. The project selected was to apply and implement an existing oil fates model (OSCAR) into the Nowcast/Forecast system. The model system does not have a biological impact model component. Thus, OSRI's selection process for grants under this program was not advertised in the solicitation. In addition, the scope changed after advertising the BAA because the simulations of the spatial-temporal distribution and impacts of dispersed and nondispersed oil spills are now described as a different project called "Dispersion Impact Analysis." The intended scope of this 2002 project (budget $150,000) is to use the OSCAR model to conduct a range of dispersion scenarios to assist in future dispersant use planning and priority setting in real spill situations. The dispersion impact analysis project has not been advertised as a BAA and OSRI plans to run this activity in-house, using its staff.

Offering extremely vague solicitations opens the door to accusations that the winning proposals could only be submitted by people with previous knowledge of specifically what was desired. Also, the funding announced as available for many of the projects is so low that only someone already in the area could reasonably do the work. OSRI must write BAAs and RFPs that are consistent with its mission and long range plan and not simply write solicitations designed to garner known responses.

The committee deliberated on the appropriateness of using BAAs as the primary funding vehicle for soliciting research proposals, in contrast to other vehicles such as an RFP process. BAAs are appropriate for a number of the activities conducted by the OSRI (e.g., many of the educational activities or when novel technologies are being solicited), but projects with highly specific outcomes—that is, solicitations seeking a specific model—would be better solicited through an RFP process.

This discussion strikes at what is, perhaps, a fundamental question about OSRI, and confusion about the answer is at least in part what causes outside scientists to have some skepticism about the fairness of the organization's procedures. Is OSRI a granting institution that allows the scientific community to drive the direction and mixture of projects (through the proposal process) under the general guidance of the mission, staff, and Advisory Board, as is stipulated in the authorizing legislation, or is OSRI management responsible for developing a directed science plan that they then implement through directed procurement of specific projects and products? OSRI is now operating in this latter mode. Both approaches have advantages and disadvantages, and neither is "more" right than the other. But it is likely that many in the science community, as well as other stakeholders, believe that the first approach is what was intended, leading to a perception of unfairness in how the program is currently administered.

If the top-down approach is a deliberate choice, OSRI needs to articulate this clearly and honestly, to avoid misunderstandings and disenchantment by proposers. And OSRI should be aware that this approach poses the risk of leading to a program that, overall, appeals primarily to the stakeholders in charge. It may be that this dilemma is behind the difference of opinion (explored in depth in Chapter 7) about the appropriate scope and focus of OSRI's modeling efforts: this component is a highlight to many OSRI decision makers, but this committee believes the current emphasis on real-time spill response is inappropriate and perhaps duplicative of other efforts in the area.

These issues need to be addressed by the OSRI Advisory Board. It should consider whether the "top-down" approach is in the spirit of the legislation, whether it is effective and fair to appear to manage a broad solicitation process when most degrees of freedom are proscribed, whether there is an adequate mechanism in place to set priorities and vet ideas, or whether the program should be under the strong direction of management.

THE PEER REVIEW PROCESS

According to OSRI, peer reviewers are selected by staff and reviewed/approved by the OSRI director. The staff focuses on academic institutions to provide peer reviews but also uses board and STC members, industry, agencies, and others to select peer reviewers with the proper expertise to review specific proposals. The February 2002 changes to the Grant Policy Manual include a provision that the STC members review all proposals recommended for award for $25,000 or over; therefore, those proposals will receive additional peer reviews, beyond the three required. People who submit proposals under a given BAA are not used to review other proposals under that BAA, and all reviewers are required to declare potential conflicts of interest and are not used if there is a conflict. When asked if there were requirements for geographical, expertise, or other criteria for the reviewer pool for an individual proposal, OSRI staff noted that reviewers are sought who possess relevant and appropriate knowledge of the subject, and that might include geographic or other specific expertise. A fairly even mixture of academia and private sector reviewers has been used. OSRI prefers to use the national and international academic arena to select peer reviewers for science proposals when appropriate.

The committee received a number of external comments, many negative, related to how proposals have been handled in the past. According to the current Grant Policy Manual, all proposals are reviewed "by at least three individual peer reviewers through a mail review and/or a panel review." In the past, proposals often had only two reviews with a third requested if "deemed necessary." This increase in number of reviewers is one of several examples of changes that have been made by the Advisory Board to improve the proposal process.

Until now all proposal reviews (with the exception of a few of the smaller education proposal reviews) have been accomplished through mail rather than panel review, although policy allows for either option. In the past, according to the OSRI director, staff formed small panels to make final funding decisions on all small and medium grants. Now, the Scientific and Technical Committee is used to decide funding on all medium size grants. The committee believes that recent changes in OSRI procedures will improve the public perception of fairness in the process, and we commend OSRI and the Advisory Board for their continued efforts to improve the process. OSRI might consider adapting additional protocols of the National Science Foundation in selection of reviewers. For example, some reviewers could be selected from names provided by applicants. As with NSF, potential applicants could have the option of requesting that one or two people not be used as reviewers. Furthermore, applicants could be requested to submit information identifying relevant close colleagues such as previous graduate studies advisors, former graduate students, and co-principal investigators on other recent projects, so that these people are not inadvertently selected as reviewers.

The broad categories assessed by reviewers include need, benefits, implementation, financial matching, experience, milestones, repayments, and risk sharing. Once reviews have been completed, proposals are grouped according to their rankings: those receiving "fund" rankings only are placed at the top of the list, those receiving any "fund after minor revisions" are ranked in the middle of the list, and those receiving "do not fund" rankings are placed at the bottom of the list. Within each of those categories, decisions to fund are based on the ranking in the list, total dollars available, and the matching funds included in the project budget.

Although there have been problems (some real and some perceived) with the proposal process used by OSRI, most of these have been addressed by recent policy changes in the way grants are processed. If present policies for peer and panel review (as well as management of conflict of interest) adapted from those used by other granting institutions such as the National Science Foundation, Office of Naval Research, Minerals Management Service, and National Oceanographic Partnership Program are followed carefully, the process should be fair.

While all proposals are required to go through the above review process, final approval for funding differs depending on the funding level. Small projects can be approved directly by the OSRI director, medium-size projects by the Scientific and Technical Committee (following recommendation by the director), and large projects by the OSRI Advisory Board (following recommendation by the director, and the Scientific and Technical Committee). When there is a conflict between the director and the Scientific and Technical Committee regarding funding recommendations for medium-size proposals, the final decision is made by the OSRI Advisory Board. The policy does not spell out the level of participation by the Scientific and Technical Committee. Our committee would urge the STC to take an active role in evaluating the proposals and reviews forwarded to them.

Special conditions apply to proposals submitted by personnel in the PWSSC. The most significant of these is that "all proposals ranked higher than a proposal submitted by the Prince William Sound Science Center (PWSSC) shall be offered an award prior to the Center's proposal(s) being approved for funding support." Should a situation occur where a PWSSC proposal is tied in ranking with a proposal from an outside entity, the proposal from the outside entity would be preferentially funded. To date, this situation apparently has not occurred. This preference to non-PWSSC proposals is a good policy, given that there have been perceptions of favoritism in the center.

NUMBER OF PROPOSALS AND SUCCESS RATIO

Responses to OSRI BAAs have ranged from 0 to 14 proposals (Table 3-1). Although the dataset is not large, there appears to be a downward trend in the number of proposals submitted for each BAA over time. For example, in FY98, 7 of 11 BAAs had more than 5 responses each, but by FY01 the greatest number of responses to any single BAA was 3. In FY00 only 3 of 9 BAAs had responses greater than 5 per BAA. Proposals actually funded for each BAA vary from 0 to 5. One item in the information provided to the committee appears incorrect: in FY01, 3 grants were made where only 1 proposal was received. All of these grants were for educational activities and were "small" grants as defined by the OSRI's guidelines.

DISTRIBUTION OF FUNDS BY CATEGORIES

As mentioned earlier, the OSRI Advisory Board mandated that funds be spent in an approximate split of 40 percent for applied technology, 40 percent for predictive ecology, and 20 percent for education and outreach. These targets were more or less met in the four years evaluated (Table 3-2).

The committee was also interested in how OSRI distributed funds relative to the size of the grant awarded. Projects were broken down into small (less than $25,000), medium size ($25,000-$99,999), and large ($100,000 and greater) grants (Table 3-3). Grants were awarded across all categories with medium-size and large proposals comprising about an equal split in funding for a given year. While current policy dictates approval of medium-size proposals by the Scientific and Technical Committee and large proposals by the Advisory Board, it is our understanding that this policy was not in place for our evaluation period (FY98-01). The distribution of funds allowed the OSRI to fund from 23-25 projects each year from FY99 to FY01. Even with the current approval process in place, the director could spend a substantial portion of the budget by approving many small grants. The board may want to consider a maximum limit per year that can be spent on small projects. However, our committee believes that the distribution of funds among small, medium-size and large grants was appropriate over the period evaluated (FY98-01).

COMPLIANCE WITH PROCEDURES

The OSRI proposal review process has evolved over time. Examples of changes in policy are the increasing the number of reviewers for each proposal and restricting the sole approval of projects by the director to those less than $25,000. The committee received a number of outside comments on the OSRI program, some of which expressed concern about what they believed to be actual or perceived fairness issues in the OSRI grant process. Some outside scientists perceived that OSRI relies too heavily on PWSSC staff for proposal review, that advertising of BAAs is not timely and some people are given the "inside track" before formal announcements are made, that deadlines set in BAAs are sometimes inappropriately short, and other issues. Some of these concerns have been addressed by official policy changes and some (whether real or perceived) could be addressed by formalizing the advertising of BAAs and RFPs and making advertisement of both more widespread.

RELATIONSHIP BETWEEN PWSSC AND OSRI

The historical relationship between the Prince William Sound Science Center (PWSSC) and the Oil Spill Recovery Institute (OSRI) is expressed in the document "A Brief Overview of Prince William Sound Oil Spill Recovery Institute (OSRI) and Science Center (PWSSC)," which states,

> In 1989, the Prince William Sound Science Center (PWSSC), a private 501(c)(3), non-profit corporation, developed and pursued the authorization of the Oil Spill Recovery Institute (OSRI) in the Oil Pollution Act of 1990 (OPA90) with assistance from the City of Cordova. OSRI was designed to be administered by PWSSC while its program is approved by an advisory board with strong public, industry and Alaska Native representation and a balance of relevant agency and academic members. Congress appropriated funds for the OSRI program in the Coast Guard Authorization Act of 1996 (CGAA96), after six years of research and education programs had established a credible science program at the PWSSC.

OSRI is administered by the PWSSC and shares administrative support with the PWSSC, perhaps most notably a director. By law, however, the OSRI has a separate mission to "identify and develop methods to deal with oil spills in the Arctic and subarctic environment and work to better understand the long-range effects of oil spills on the natural resources of Prince William Sound and its adjacent waters...." OSRI has come under

TABLE 3-1   Number of Proposals Submitted and Funded in Response to Broad Area Announcements (BAAs) by Fiscal Year

Year Announced	BAA Title	Program Area
FY98	Ice Hazards Workshop	Public Education & Outreach
	Technical Writer	Public Education & Outreach
	Small Spill Education	Public Education & Outreach
	Webmaster	Public Education & Outreach
	Public Information Services	Public Education & Outreach
	Dispersant Workshop	Public Education & Outreach
	Community Education	Public Education & Outreach
	Nowcast/Forecast Ocean Circulation	Applied Technology & Predictive Ecology
	Workshops (standing solicitation)	Public Education & Outreach
	Student Internships (standing solicitation)	Public Education & Outreach
	Graduate Fellowship (standing solicitation)	Public Education & Outreach
FY99	Graduate Fellowship (standing solicitation)	Public Education & Outreach
	Ecology Programs	Predictive Ecology
	Workshops (standing solicitation)	Public Education & Outreach
	Student Internships (standing solicitation)	Public Education & Outreach
FY00	Remote Sensing	Applied Technology
	Dispersed vs Nondispersed Simulation	Applied Technology
	Small Spill Technology	Applied Technology
	Webmaster	Public Education & Outreach
	Plankton Monitoring	Predictive Ecology
	Herring & Pollock Monitoring	Predictive Ecology
	Graduate Fellowship (standing solicitation)	Public Education & Outreach
	Student Internships (standing solicitation)	Public Education & Outreach
	Workshops (standing solicitation)	Public Education & Outreach
FY01	Alaska Vessel Transponder System	Applied Technology
	Community Education (standing solicitation)	Public Education & Outreach
	Tide Height Data Collection	Applied Technology
	Meteorolgical Data Collection	Applied Technology
	Current Validation	Predictive Ecology
	Hinchinbrook Entrance ADCP	Predictive Ecology
	Graduate Fellowship (standing solicitation)	Public Education & Outreach
	Workshops (standing solicitation)	Public Education & Outreach
	Student Internships (standing solicitation)	Public Education & Outreach

NOTE: The year the BAA was announced is not always the year the project was funded. This table includes only projects funded through the BAA process. The OSRI staff provided the data for the table.

Award Category	Number Submitted	Number Funded
Small	4	1
Small	7	1
Small/Medium	8	0
Small	10	1
Small	7	1
Small	8	1
Small/Medium	14	3
Large	10	3
Small/Medium	0	0
Small	3	3
Small	2	2
Small	1	1
Medium	13	5
Small/Medium	5	4
Small	1	1
Medium	7	1
Large	7	1
Medium	2	1
Small	10	1
Large	1	1
Large	2	1
Small	3	2
Small	2	2
Medium	2	2
Small	2	1
Small	1	3
Medium	3	1
Medium	3	1
Medium	2	1
Medium	3	0
Small	3	2
Small/Medium	2	1
Small	1	1

TABLE 3-2   Amount of Funds Spent FY98-01 by Category

| Activity | Amount (Percent of Total) | | | |
	FY98	FY99	FY00	FY01
Applied Technology	$47,767 (24%)	$397,753 (39%)	$449,876 (36%)	$511,099 (34%)
Predictive Ecology	$29,879 (15%)	$451,337 (45%)	$520,472 (42%)	$580,000 (38%)
Education and Outreach	$97,045 (49%)	$63,069 (6%)	$175,018 (14%)	$206,169 (13%)
Technology Coordinator	$13,598 (7%)	$100,000 (10%)	$100,000 (8%)	$121,175 (8%)
Other	$8,943 (5%)	$0	$0	$100,000[a] (7%)
TOTAL	$197,232	$1,012,159	$1,245,366	$1,518,443

NOTE: The numbers were calculated from data provided by the OSRI, "OSRI Contracts Awarded 1997-Present," and are based on amount spent.

[a]National Academy of Sciences review.

TABLE 3-3   Breakdown by Year of Projects Funded as Small, Medium-Size or Large Awards

| Type of Award | Total Dollar Amount (Number of Grants) | | | |
	FY98	FY99	FY00	FY01[a]
Small	$124,634 (12)	$137,678 (12)	$91,329 (8)	$137,826 (9)
Medium-Size	$59,000 (2)	$374,750 (8)	$593,037 (12)	$659,442 (12)
Large	$0 (0)	$399,731 (3)	$461,000 (3)	$500,000 (4)
TOTAL	$183,634 (14)	$912,159 (23)	$1,145,366 (23)	$1,297,268 (25)

NOTE: Categories based on amount awarded. The technology coordinator position was not considered an award in this table. In calculating numbers for the table, awards of $25,000 or less were considered small, awards of $25,001 to $99,999 were considered medium-size, and awards greater than or equal to $100,000 were considered large.

[a]National Academy of Sciences review was not included.

some criticism relative to its close links with the PWSSC. Specifically, concern has been raised whether PWSSC researchers have received a disproportionate amount of funding from the OSRI. A breakdown of support to PWSSC researchers from the OSRI program is summarized in Table 3-2. Data used to generate Table 3-2 were obtained from the tables "OSRI Contracts Awarded 1997-Present" provided by the OSRI.

In examining Table 3-4, about 30 percent of the total dollars awarded by the OSRI have been granted to PWSSC projects each year since FY98. The most significant support was to projects categorized under predictive ecology. OSRI records indicate that 70 percent of the funds spent in predictive ecology in FY00 and 66 percent in FY01 were awarded to PWSSC projects. In contrast, with the exception of FY99, little support has been given to PWSSC projects in the area of applied technology. Support of PWSSC projects in the area of education and outreach has steadily decreased (as a percent of total support) since FY98.

Another way to look at the fiscal relationship with respect to awards between the OSRI and the PWSSC is to look at the number of proposals awarded to PWSSC projects relative to the total funded (Tables 3-4 and 3-5). From 28 percent to 40 percent of the total projects funded were awarded to PWSSC projects. This result (about 30 percent to PWSSC each year) is in line with the total dollar amounts awarded (Table 3-3) and is not unreasonable.

TABLE 3-4   Amount of Total Funding by Category Awarded to Prince William Sound Science Center

| Program | Amount (Percent of Total Amount in Category from Table 3-1[a]) | | | |
	FY98	FY99	FY00	FY01
Applied Technology	$8,000 (17%)	$133,100 (33%)	$0 (0%)	$0 (0%)
Predictive Ecology	$0 (0%)	$212,627 (47%)	$365,215 (70%)	$380,215 (66%)
Education and Outreach	$34,883 (36%)	$18,348 (29%)	$35,091 (20%)	$28,000 (13%)
Other	$8,943 (100%)			
TOTAL	$51,826 (28%)	$364,075 (40%)	$400,306 (35%)	$408,215 (29%)

[a]The technology coordinator position was not included in the total for this table.

TABLE 3-5   Breakdown of Funding Awarded to Prince William Sound
Science Center Projects by Category or Project and Funding Year

Type of Award	Number Awarded to PWSSC/Total number awarded (from Table 3-2)			
	FY98	FY99	FY00	FY01
Small	4/12	3/12	2/8	0/9
Medium	1/2	1/8	4/12	4/12
Large	0/0	2/3	1/3	1/4
TOTAL	5/14	6/23	7/23	5/25

NOTE: See Table 3-2 for total numbers of projects in each category.

PUBLICATIONS BY PRINCIPAL INVESTIGATORS

The committee evaluated a list of publications provided by OSRI (Appendix H). The publication list includes publications by principal investigators sponsored by OSRI and various workshop proceedings and other documents. The list includes 44 total publications described as published or "in press." Of these 44 publications, 15 are journal articles, 15 are conference proceedings, 6 are abstracts, 4 are videos, and 5 are various maps, guides, or other reports. Of the journal articles, one by Thomas and Thorne (2001) was published in *Nature.* Four journal articles were published in 2000, one in 2001, and 10 are listed as in press. For overall publications, 5 were listed in 1998, 5 in 1999, 8 in 2000, 15 in 2001, 1 in 2002 (to date), and 10 in press (as of August 2002).

When first evaluating the list provided by OSRI in early 2002, the committee felt that the publication rate, particularly in refereed journals, was lower than it should be for a program the size of OSRI, because at that time it appeared that only 5 items had been published in peer-reviewed journals. Currently, however, 10 other journal articles are listed as in press, indicating that OSRI-sponsored research is starting to make its way into the refereed literature. The committee strongly encourages OSRI to emphasize the funding of projects that are likely to lead to information that can be published in refereed journals with broad distribution. OSRI should also take care to be sure that it is credited for all the publications and presentations that arise from the projects it supports (e.g., fellowships often result in theses and publications that often go unreported).

4

Program Planning

The charge to the committee included a request to consider whether existing documents set an appropriate course for the future. To this end the committee was provided with a number of documents, including

- Oil Spill Recovery Institute (OSRI) Bylaws
- Public Law 110-380-Aug. 17, 1990, Title V. Prince William Sound Provisions
- Oil Pollution and Technology Plan for the Arctic and Sub-Arctic, September 1995
- Business Plan OSRI April 1999
- OSRI Annual Work Plan FY00
- OSRI Annual Work Plan FY01
- OSRI Annual Work Plan FY02

Neither the OSRI bylaws nor the enabling legislation (Public Law 110-380-Aug. 17, 1990) includes a requirement for a strategic plan for OSRI or guidance for review and implementation of a strategic plan. The 1995 Oil Pollution and Technology Plan for the Arctic and Sub-Arctic is the *de facto* initial strategic plan for OSRI. Subsequent OSRI annual work plans are actually tactical planning documents. The overall objectives of the OSRI program are restated with variations and changes in emphasis in each plan. The linkage between the 1995 strategic plan and the annual tactical plans is often vague and it is difficult to track the evolution of strategic planning for the overall OSRI program.

The 1999 business plan contains a section called "Strategic Vision" that provides the rationale and justification for the focus of the program on "physical and biological prediction," the Nowcast/Forecast (NC/FC) program. Weakness in physical and biological prediction tools is portrayed as a major limitation to improving oil pollution prevention and response. The entire OSRI program, including education programs and technology development, is meant to focus on this strategic vision.

The committee considered available OSRI annual reports 1997-1999, 1999, and 2000 and the OSRI FY02 Technology Coordinator's Report to assist in assessing whether the strategic plans were being implemented. The director identified a series of workshop reports used for planning purposes:

- Thomas and Cox (2000), AMOP, "A Nowcast/Forecast System for PWS;"
- Robertson and DeCola (2001), Proceedings of the "PWS Meteorological Workshop;" and
- Dickens (2002), "Oil and Ice R&D Priorities."

As can be seen from the documents provided, OSRI planning is based on a range of documents including specific planning documents, annual plans, and workshop reports. In general, these documents review and restate the implementing legislation and the OSRI mission and goals. The first long-range (10 years) planning document outlines the plan for implementing OSRI and identifying the objectives of OSRI as

> The goal of OSRI's research and development is to identify technologies (hardware, software and procedures) that reduce the risk of an oil spill, and risk of damage to the environment. (OSRI, 1995)

This is to be accomplished by a three pronged approach: (1) reducing the risk of oil spills—prevention; (2) reducing the risk of damage after a spill—response research; and (3) assessment of the ecosystems along the transportation route—ecosystem assessment. The report then goes on to provide a list of recommendations that set priorities based on gaps in information: Forty-seven recommendations are made. Overall, the document is well written and provides adequate guidance for implementing the OSRI mandate.

The larger question, in addition to the adequacy of planning documents, is whether the plan is being implemented and whether progress is being made toward the stated goals. It appears that the annual plans are intended to answer these questions. The oldest work plan provided to the committee was a short business plan dated April 1999. An initial business plan was developed in 1997 and this is when the 40/40/20 percent fund distribution goal was established. The annual plans summarize the background information on the formation and goals of OSRI. Little explanation for the changes in emphasis or goals was found in the documents provided. The 1995 Oil Pollution and Technology Plan for the Arctic and Sub-Arctic is not mentioned. Part of the 1999 business plan is a section entitled "A Strategic Vision," and this outlines a new set of goals, approaches, and objectives. The strong emphasis on predictive ecology is set forth as a principle in these documents. Over time, the annual plans become primarily a listing of funded programs with little cross-program synthesis or explanation of how each plan addresses the mission and goals of OSRI.

The next type of document is entitled the *Technology Coordinator's Report* (2002) and is called a status report to the OSRI Advisory Board. This report reviews the finances and then provides a program-by-program summary fact sheet. The fact sheet includes a summary of the program including budget and partners/cost share, a timeline, end users, relationship to OSRI mission, details of proposal award process, and any updates. NC/FC is defined as the principle development effort of OSRI. While a section is included on relevance to the OSRI mission, the connections are not explicit in most cases and relevance is often justified on nonmission criteria. For example, certain projects are identified "to aid in tactical operations related to oil spill containment and mitigation." This report is a useful compilation of information regarding OSRI-funded activities. The report could be improved by more discussion on how the projects fit together into a coordinated program of research and technology development.

There has been an evolution in the type and content of planning documents since the creation of OSRI. In general, the documents have been improving in quality and provide useful information to oversight groups. In particular, the technology coordinator report is helpful in providing a comprehensive picture of what OSRI does. These reports could be improved by adding a synthesis chapter tying the projects together into an overall vision.

There is a need for a revised strategic plan. The strategic plan should be used to inform decisions regarding all OSRI actions. The question to be asked is: How does each activity relate to the OSRI mission? This is essential to ensure that sound and consistent decisions are made over a period of years. Strict adherence to mission will provide a sense of fairness and equity even when difficult and controversial decisions must be made. Arbitrariness will be inferred if the strategic relevance and justification of decisions is not made clear. This can only occur within the context of an explicit, clear, and detailed strategic plan that guides the daily operations of OSRI.

According to the strategic vision statement in the 1999 business plan, all OSRI programs are intended to help improve our ability to predict marine ecosystem changes and the centerpiece of the OSRI program is the NC/FC program. However, it is unclear in most cases how many of the projects sponsored by OSRI contribute to this strategic goal. The acoustic monitoring of fish and zooplankton populations is providing valuable ecological data; however, it is unclear how the snapshot of herring, pollock, and zooplankton distributions and abundance during a short time period each year and in a confined geographic area of Prince William Sound (compared to the geographic range of the populations being monitored) can be used as input to the Nowcast/Forecast and OSCAR model systems. The Copper River Delta ecology program is good science but how will the results fit into he overall OSRI program? A strategic plan should clearly identify the linkages among all program elements. It also should be updated periodically to help refocus the program as new data or requirements dictate.

As OSRI plans its future activities, it must take steps to tie its planning into what other research organizations are doing, or at minimum plan its activities with more awareness of what other organizations are doing. Environment Canada, for example, has six committees that plan and review research, and these meet annually each December to plan future work and establish connections with other groups. As many as 35 other agencies in North America and Europe are represented. Similarly the American Petroleum Institute (API) has a Spill Advisory Group that looks at API research and includes representatives of about 10 organizations. Both of these are opportunities for OSRI to tie its program planning into existing coordination efforts.

5

Predictive Ecology

One of the more profound outcomes of the 1989 *Exxon Valdez* oil spill was the recognition of our limited ability to realistically predict the effects of an oil spill on marine resources. The ongoing debate over long-term damages further highlights just how inadequate previous knowledge was in attempting to discern cause and effect in natural environments. This lack of knowledge was, on one level, an incomplete understanding of what resources were present. But even more fundamental was a lack of understanding of the structure and functioning of complex ecosystems. Our knowledge of the interconnectedness of systems and the basic controls on biological patterns in time and space was wholly inadequate to answer the questions being posed by regulators, litigators, responders, stakeholders, and the public in general. Without this fundamental knowledge, predictions, prevention, response, remediation, and resource damage assessment can be in jeopardy of being ineffective or inaccurate and thus potentially wasting financial resources and putting natural resources at risk.

One component of the OSRI mission is the portfolio of projects funded in the category called "predictive ecology" (Table 5-1). To facilitate the committee's review of the OSRI portfolio, it was convenient to further subdivide predictive ecology into its modeling and nonmodeling efforts. This is in part due to the large effort being expended to develop the Nowcast/Forecast (NC/FC) model, which is a component of both the predictive ecology and applied technology programs. While many of the nonmodeling activities may support the modeling efforts, in many instances they are more or less stand-alone projects, and the committee chose to

TABLE 5-1   Summary of OSRI-Funded Projects FY98-02 by Program
Area: Predictive Ecology

Contract Term	Project Title[a]	Total[b]
06/15/98 - 12/15/98	Internship: Waterfowl Toxicity Study	$6,000
07/01/98 - 06/30/99	Fellowship	$23,879
11/20/98 - 03/30/99	Science Planning: Sound Science Research Team	$7,363
01/01/99 - 12/31/99	Nowcast/Forecast Program	$133,298
04/15/99 - 04/14/00	Statistical Methods & Software	$26,566
06/01/99 - 05/31/00	Sentinel Rock Fish Monitoring	$71,966
05/01/99 - 04/30/00	3-D Coupled Biological-Physical Model for the Ecosystem in Prince William Sound	$49,974
05/01/99 - 04/30/00	Responses of River Otters to Oil Contamination: Monitoring Post-Release Survival of Oiled and Nonoiled Captive Otters.	$24,100
05/01/99 - 04/30/00	Workshop: Cook Inlet Oceanography	$31,575
06/01/99 - 05/31/00	Waterfowl Toxicity—Measure CYPIA Induction	$50,000
06/14/99 - 06/13/00	Environmental Sensitivity Maps: Prince William Sound	$50,000
09/01/99 - 02/28/00	Workshop: Cook Inlet Safety of Navigation	$11,826
01/20/00 - 01/19/01	Herring/Pollock Monitoring in Prince William Sound	$75,000
01/20/00 - 01/19/01	Zooplankton/Nekton Monitoring in Prince William Sound	$60,000
02/01/00 - 01/31/01	Nowcast/Forecast Program	$150,000
04/01/00 - 03/31/01	Distribution/Abundance of Intertidal Invertebrates on the Copper River Delta	$80,215
04/01/00 - 03/31/01	Distribution/Abundance of Intertidal Invertebrates on the Copper River Delta	$19,785
06/28/00 - 07/27/01	Environmental Sensitivity Maps: Aleutian Islands	$85,472
	3-D Coupled Biological-Physical Model for Prince William Sound	$50,000
	Sensitivity Mapping Project: Southeast Alaska	$20,000
	MOU—Environmental Sensitivity Mapping	$60,000
	MOU—Remote Sensing Using Lidar	$100,000
10/01/00 - 09/30/01	Technology Coordinator[c]	$40,391

Program Area	Institution Awarded	Modeling/Support of Modeling?
Eco	University of California, Davis	
Eco	University of Alaska, Fairbanks	
Eco	Prince William Sound Aquaculture Corporation	
Eco	Prince William Sound Science Center	✓
Eco	H.T. Harvey & Associates	
Eco	Prince William Sound Science Center	
Eco	University of Alaska, Fairbanks, Institute of Marine Sciences	✓
Eco	University of Alaska, Fairbanks/IARC	
Eco	University of Alaska/SFOS	
Eco	University of California, Davis	
Eco	NOAA/NOS/OR&R	
Eco	Cook Inlet RCAC	
Eco	Prince William Sound Science Center	
Eco	Prince William Sound Science Center	
Eco	Prince William Sound Science Center	✓
Eco	Prince William Sound Science Center	
Eco	University of North Carolina	
Eco	Research Planning, Inc.	
Eco	University of Alaska, Fairbanks/IARC	✓
Eco	SEAPRO	
Eco	NOAA/NOS/OR&R	
Eco	NOAA/Remote Sensing Lab, Colorado	
Tech/Eco/Edu	OSRI	

continued

TABLE 5-1   Continued

Contract Term	Project Title[a]	Total[b]
02/01/00 - 01/31/02	Herring/Pollock Monitoring in Prince William Sound	$75,000
02/01/00 - 01/31/02	Zooplankton/Nekton Monitoring in Prince William Sound	$75,000
02/01/00 - 01/31/02	Nowcast/Forecast Program	$150,000
04/01/01 - 03/31/02	Distribution/Abundance of Intertidal Invertebrates on the Copper River Delta	$80,215
04/01/01 - 03/31/02	Distribution/Abundance of Intertidal Invertebrates on the Copper River Delta	$19,785
TOTAL		$1,627,410

NOTE: All totals are approximate and are based on information provided by OSRI in February 2002.
[a]Descriptions of most projects can be found at the OSRI website <http://www.pwssc-osri.org>.

evaluate the modeling and nonmodeling activities separately (see Chapter 7).

Within the predictive ecology program (nonmodeling portion), OSRI has funded a diverse set of projects, including workshops, fellowships, a study of the toxic effects of oil on waterfowl, monitoring of rockfish, monitoring of river otters exposed to oil, herring and pollock monitoring, study of the distribution of intertidal invertebrates on the Copper River Delta, sensitivity mapping of coastal resources, and zooplankton/nekton monitoring (Table 5-1). Excluding the modeling components that OSRI categorizes within predictive ecology, funding of these nonmodeling projects totals a little more than $1 million in the FY98-01 period.

SELECTED PROJECT DESCRIPTIONS

To gain a sense of the direction of and priorities within the Predictive Ecology program (nonmodeling components), the committee looked at the program's largest projects: herring, pollock, zooplankton, and nekton monitoring in Prince William Sound; the Copper River ecological study; sensitivity mapping of coastal resources (in partnership with NOAA); and a continuation of the herring and pollock acoustic monitoring activity.

Program Area	Institution Awarded	Modeling/Support of Modeling?
Eco	Prince William Sound Science Center	
Eco	Prince William Sound Science Center	
Eco	Prince William Sound Science Center	✓
Eco	Prince William Sound Science Center	
Eco	University of North Carolina	

*b*This is the total spent on the project through early 2002; some projects continue.

*c*As of this entry, the technology coordinator is listed in all three programs (ecology, technology, and education), rather than just in technology, so Tables 5-1, 6-1, and 8-1 each include one-third of $121,175 in their totals.

Resource Monitoring in Prince William Sound

The Prince William Sound ecological resources monitoring projects focus on the three dominant pelagic animal biomass (most abundant) animals in the sound. These are herring, pollock, and copepods of the genus *Neocalanus*. Herring monitoring began in 1993 (pre-OSRI), pollock monitoring in 1995, and zooplankton monitoring in 2000. Acoustic and optical sensing techniques are used to monitor seasonal distributions and abundances.

Together the monitoring projects will provide information about the target animal populations and, to a lesser extent, their roles in the Prince William Sound ecosystem. Some aspects of the research have been published (Thomas and Thorne, 2001). However, the optimization strategy used in designing these programs has raised some concerns. Prince William Sound is a large embayment, and the winter and spring distributions of herring and pollock may not be fully identified by the current field design. In the past, fisheries biologists have suggested that juvenile herring from the sound may over-winter in the bays and coastal water of the Kenai Peninsula. Early studies of herring (1930s), when the main fishery was for adult herring for pickling, show a very different distribution than most fisheries studies show today. Similarly, pollock stocks have shown dramatic shifts in biomass and distributions in the Gulf of Alaska, Prince William Sound, and the Bering Sea in recent years as more valuable fish stocks have been over-harvested. Climate change also is likely to have played a role, primarily through changing water temperature and related changes in primary production.

The dynamics of zooplankton populations are extremely complex. The acoustic survey methods in this project provide information about their abundance during the spring bloom period. Research has demonstrated significant interannual variability in zooplankton biomass and species composition in Prince William Sound (Cooney et al., 2001; Eslinger et al., 2001) and geographic variation in the bloom locations likely contributes to these differences. Given the current sampling design, it may be difficult to use the results of zooplankton studies in the spring to model spatial and temporal trends of large zooplankton species in the biological modules of the NC/FC model system.

The monitoring projects will provide valuable information about animal distributions during critical high aggregation times. This work should continue to evaluate the seasonal distributions and abundances of high biomass species. The stated objective is that the results of this monitoring will be integrated into the NC/FC model. However, at this point biological modeling components are not being developed.

Challenges remain in confirming the identity of targets in acoustic signals and calibrating the instruments accurately to biomass or other important indicators of stocks, such as size and composition (i.e., age distribution and sex). Denser geographic and temporal coverage is needed to fully understand the dynamics of these biomass dominants in Prince William Sound and adjacent areas. If the results are to be effectively integrated into biophysical models, supporting chemical and physical measurements, such as nutrients, particulates, chlorophyll, and oxygen, will have to be collected as well. Closer alignment of these studies with regional fisheries assessments and studies of other populations are important. The objectives of these projects are laudable, but full implementation may take years and exceed the financial wherewithal of OSRI. Leveraging, partnering, and collaboration with larger ecosystem programs in the area will be crucial to developing a truly predictive model of these important and complex ecosystems.

Copper River Delta Study

A major OSRI predictive ecology project is a study of the lower Copper River Delta ecosystem. The research focuses on the distribution of bivalves, amphipods, and insect larvae. The study area and target species appear to have been chosen based on their potentially high sensitivity to petroleum hydrocarbons and their importance to the diet of shorebirds frequenting the area. The Copper River Delta is an important seasonal habitat for migratory birds using the Alaskan coastal flyway. This study provides basic ecological information about a unique habitat. There have been few other studies of the ecology of river estuary systems in subarctic environments, such as the Copper River estuary. Such ecosystems, including the Upper Cook Inlet, are oligotrophic (high biomass and low diversity), due mainly to climatic conditions and high sediment loads derived from glacial till, and usually are not particularly sensitive to the organic enrichment stress produced by an oil spill. However, they are quite productive and, particularly in the case of the Copper River Delta, are known to be important foraging areas for migratory birds.

This project is well designed and will provide valuable information about this type of delta habitat. However, as currently designed, the project does not address some questions whose answers would be needed to respond to or assess the impacts of a future oil spill in the larger area. For example, there is no chemistry component for evaluating existing polynuclear aromatic hydrocarbon (PAH) sediment concentrations or PAH source characterization in the existing program.

Although the Copper River ecosystem is clearly of ecological and economic importance, especially to the community of Cordova, the ecosystem is not representative of other Arctic and subarctic biomes and thus, it may prove difficult to extrapolate the findings to other settings. In the future, representativeness of Arctic and subarctic environments should be an important consideration when choosing sites for long-term monitoring and study.

Environmental Sensitivity Index

OSRI has provided significant support ($215,000) to a project conducting environmental sensitivity index mapping for Alaskan coastlines. The project supports standard mapping of resources in Prince William Sound and the Aleutian Islands, focusing on coastal regions so responders would be better able to protect sensitive areas in the event of a spill. The Hazardous Materials Response Division of NOAA's Office of Response and Restoration (HAZMAT), which coordinates advice on science and natural resource issues for the US Coast Guard, and which is charged with directing spill response under OPA 90. The mapping is used by HAZMAT in identifying resources at risk from a spill. The focus of the mapping is on shoreline type and degree of exposure, which determines its sensitivity to oiling. Shoreline protection and cleanup priorities are set using this, as well as other information. It also provides basic background information on shore types useful in natural resource damage assessments. HAZMAT has been developing such data for the entire U.S. shoreline, and thus the OSRI funds are essentially a supplement to that effort. This project is within the OSRI mission.

Remote Sensing Cooperative Agreement

The LIDAR Proof of Concept project (also called Remote Sensing Cooperative Agreement) is a one-year project that builds on the herring and pollock acoustic monitoring program initiated in FY00. LIDAR in conjunction with established acoustic methods represents a potential for precise estimates of these key fisheries. Funding within this project will enable a proof-of-concept effort to establish the viability of this technique. This project was allocated $100,000 in FY00.

RESPONSIVENESS TO MISSION

In general, the OSRI predictive ecology projects (nonmodeling) are responsive to the OSRI mission. However, some projects are less clearly connected to the OSRI mission. One possible concern is the OSRI emphasis on Prince William Sound, when the legislative mandate is geographically much broader. The committee understands and supports the need to focus on the resources most at risk from oil spills, given the limited financial resources. This would be mainly Prince William Sound and possibly Cook Inlet (particularly if oil exploration and development increases again).

The legislative mandate implies that studies of long-term ecological effects of oil spills are appropriate projects for OSRI, although to date this has not been an area of emphasis. The legislation tasks OSRI with assisting the *Exxon Valdez* oil spill trustees in assessing environmental damages from the *Exxon Valdez* oil spill, focused geographically on the northern Gulf of Alaska region, primarily Prince William Sound. There remains great potential for research on long-term ecological effects.

OSRI has given little attention to the second half of the legislative mandate "... to understand the long range effects of Arctic and subarctic oil spills on...the economy, and the lifestyle and well-being of the people who are dependent on them...." A limited amount of OSRI research is directed at the human occupants of the area and how they interact with resources and how oil spills influence these interactions. The legislative mandate is extremely broad and OSRI has limited resources, so this decision to focus is understandable.

FUTURE DIRECTIONS

There is still great need for improved understanding and targeted monitoring of sensitive areas at risk from oil spills or chronic releases of petroleum (NRC, 2002). In particular, there is a continuing need for a better understanding of the physical and biological effects of oil spills in Arctic and subarctic environments. Such information is essential for developing optimal spill prevention, response, and remediation strategies that limit collateral harm to the environment. In general, knowledge of biological resources and how to protect them is still limited. There is a great need for research to improve methods and strategies for bioremediation of oil-contaminated Arctic and subarctic marine and wetland ecosystems. There also is a need to better understand the physical, chemical, and biological fates (weathering) of petroleum in cold environments. Knowledge of biological resources, and how to protect them, is frequently lacking; marine mammals and birds offshore in the Beaufort and Chukchi seas are at particular risk. The following suggestions are illustrations of possible directions and are not intended as a comprehensive list or to replace strategic planning by OSRI.

Ecosystem Structure and Function

One of the stated objectives of the ecological studies supported by OSRI is to develop "baseline" data about the structure and function of marine ecosystems in the northern Gulf of Alaska that are at risk from future oil spills. The baseline data are intended to provide the basis for documenting and quantifying ecological injury, should a spill occur. Baseline inventories of resources or benchmarks have real limitations in preparing for and assessing damage after an oil spill. A National Research Council review of a Bureau of Land Management (BLM) program determined that the baseline approach is not a useful way to characterize the effects of human perturbations on natural systems (NRC, 1978). A fundamental restructuring of BLM's environmental studies program was recommended. The NRC conclusions are applicable to OSRI and how it selects its portfolio of projects. Studies of natural resources need to provide a dynamic vision of the systems being studied, because they change on a wide variety of spatial and temporal scales. Inventories or descriptive studies should be performed only within the context of understanding the structure and functioning of ecosystems and not as a static picture of resources at one point in time. Efforts to develop techniques for monitoring populations and implementing those techniques over multiple years will provide a more fruitful approach to understanding resources.

Collateral Damage Caused by Clean-Up Techniques

There is considerable debate, and little resolution, concerning the extent to which the methods used in oil spill response themselves cause injury to the shoreline and coastal marine resources, beyond the effects of the oil itself. Research is needed on the effects of different oil spill response options, such as use of chemical dispersants, beach cleaners, high-pressure or hot water washing, bioremediation, etc., on water column and intertidal biological communities. There is a need to develop methods for removing oil from the environment efficiently while minimizing adverse effects on marine organisms and ecosystems. Because the window of opportunity (length of time after a spill when dispersants are effective) for dispersant use is so narrow, it is essential that they be pre-approved for use in areas where a spill might occur. However, that would require careful studies to define the parameters that control the window of opportunity for dispersant use for a particular crude oil under a set of "typical" environmental conditions. A systematic and robust net environmental benefit analysis for dispersant use versus non-use would provide decision makers with the information needed to make effective use of dispersants.

Mechanistic Studies of Biodegradation

Biodegradation played an important role in the removal of oil from the water column and shore of the sound and Gulf of Alaska following the *Exxon Valdez* spill. The estimate (Wolfe et al., 1994) that 50 percent of the spilled oil biodegraded is very approximate, but probably in the ballpark. The bioremediation (enhanced biodegradation) studies had mixed results, but did prove the concept that enhanced biodegradation can contribute to removal of oil from the shore. At the time of the spill, most spill response experts felt that bioremediation would not work in a subarctic environment like the Gulf of Alaska, because ambient temperatures are too low. However Bragg et al. (1994), Atlas (1995), Lindstrom and Braddock (2002), and others have shown that bioremediation works in the Alaskan marine environment. More research is needed to better understand the chemical and environmental factors controlling natural and enhanced biodegradation of petroleum.

When assessing active treatment strategies, such as dispersant application, it is critical to have a clear understanding of the effects these treatments will likely have on biodegradation processes, because biodegradation is an important process leading to removal of contaminants from the environment. The results of existing studies examining the effects of commercial dispersants on biodegradation are specific to the product used, with some dispersants increasing and some diminishing biodegradation. Dispersants can potentially affect the relative degradation of different classes of hydrocarbons. Foght and Westlake (1982), for example, found that Corexit 9527 affected alkane degraders differently than aromatic degraders, depending on the nutrient regime. Overall these types of studies make clear that understanding naturally occurring biodegradation of petroleum hydrocarbons in the environment is of interest and the potential effects of treatment strategies are very important to appropriately decide on a response to an oil spill. Predictive models also need much more accurate information on the toxicity of hydrocarbons to a wide range of species inhabiting coastal areas.

Natural Resource Damage Assessment, Remediation, and Restoration

There is also a need to develop methods for restoring oil-impacted marine ecosystems and for facilitating natural recovery of impacted ecosystems, in accordance with the current NRDA regulations. Most marine resource populations and their supporting ecosystems begin recovery as soon as concentrations of oil decline to levels that are not toxic or otherwise harmful to populations. If large, long-lived species, such as marine birds and mammals, are affected, recovery of their populations may be slow. Human efforts to facilitate recovery probably should focus on habitat cleanup and protection. Similarly, intertidal communities of plants and animals often are severely harmed by oil washing ashore. However, the injured populations quickly recolonize affected habitats when the oil concentration and toxicity decline. The most effective remediation/restoration strategy would be to remove oil from the spill zone as quickly as possible.

OSRI should consider developing an oil spill restoration program, perhaps in collaboration with industry, the Environmental Protection Agency, or NOAA, to develop and evaluate methods for removing oil from the sea surface and the shoreline that cause minimal additional harm to biological resources.

6

Applied Technology

A rctic environments are faced with many issues related to the release of oil into marine environments. The presence of ice limits the effectiveness of most response technologies and methodologies and an oil spill in ice-infested waters could not be effectively responded to given the present state of technology. This deficiency puts many important marine natural resources at risk. Although much research has been directed at perfecting and improving oil spill response in Alaskan waters over the past 20 years, many problems remain unsolved. Challenges range from overcoming the difficulties of containing and recovering oil during ice freeze-up and breakup periods to improving or inventing technologies for tracking, containing, and recovering oil during the fast ice season. It is still unclear which, if any, of the existing techniques are effective in mitigating the effects of oil in the water and in or under the ice given Arctic environmental conditions and the resources at risk.

One component of the mission of the Oil Spill Recovery Institute (OSRI) is to "...identify and develop the best available techniques, equipment, and materials for dealing with oil spills in Arctic and subarctic marine environment...." To accomplish this component of the OSRI mission, a portfolio of projects has been funded to conduct research and carry out demonstration projects that are called "applied technology" (Table 6-1). Goals for the OSRI Applied Technology program include development of (1) tools to improve prevention and response to oil spills; (2) development of improved cleanup technologies; and (3) creation of models to inform the deployment of equipment and personnel during an oil spill response for maximum effect and mitigation. A range of activities and projects have been funded under the OSRI applied technology mandate, including workshops, portions of the Nowcast/Forecast model, Mechanical Oil Recovery in Infested Ice Waters (MORICE), development of an in situ hydrocarbon monitor, a computer simulation of dispersant application during an oil spill, an inventory of oil response equipment, and a study of radar as an ice detection method (Table 6-1). The funds for these projects total more than $1.8 million in the period FY98-01.

SELECTED PROJECT DESCRIPTIONS

OSRI projects under the applied technology component have been generally funded as stand-alone projects. Together they represent a mosaic of various types of activities and projects that can be broadly interpreted as supporting the OSRI mission. Examples of projects in this category include

Small Spill Technology—An information program for small boat harbors in PWS and Cook Inlet regions. It provides information, posters, and stickers to educate boaters on the different ways they can minimize pollution to the harbors and surrounding waters.

Scoping Initiative for Cook Inlet—Scoping to determine whether to do a future risk assessment exercise for Cook Inlet. If done, this would include a risk assessment of natural resources at risk of an oil spill.

PAH Field Monitor Development—Development of a reagentless polycyclic aromatic hydrocarbon (PAH) analyzer for field use. The instrument prototype was built and was to be field tested in Cordova in the summer of 2002 with assistance from NMFS personnel from Auke Bay.

Three Dimensional Oil Spill Dispersal Simulation—A Norwegian spill trajectory model, Oil Spill Contingency and Response (OSCAR), has been modified for use in PWS. It combines a fates model, weathering model, and strategic response model.

Ice Detection in Prince William Sound—Cooperative participation in a program to place a radar station on Reef Island for the detection of ice and ships moving in PWS. Total project cost was $847,000, with OSRI portion being $100,000.

PWS Tide Height Data Collection—Establishes automated tide gauges in PWS. Estimated cost is $50,000 for installations and an OSRI

TABLE 6-1   Summary of OSRI-funded Projects, FY98-02, by
Program Area: Applied Technology

Contract Term	Project Title[a]	Total[b]
12/29/98 - 03/30/98	Proceedings: A Symposium on Practical Ice Observation in CI & PWS	$5,261
03/09/98 - 05/15/98	Dispersant Application in Alaska -1998 Workshop Proceedings	$12,887
06/08/98 - 09/07/98	Internship: Network Administration	$8,000
12/01/98 - 04/26/99	Small Spill Workshop: "Plug the Leaks"	$21,619
	Technology Coordinator	$13,598
10/05/98 - 03/15/99	Workshop: "Geographic Response Planning"	$11,900
10/01/98 - 09/30/99	Fellowship	$36,000
01/01/99 - 12/31/99	Nowcast/Forecast Program	$133,100
01/01/99 - 12/31/99	Nowcast/Forecast Program	$133,333
03/31/99 - 06/15/99	Publication: "Field Guide for Oil Response in the Arctic"	$20,315
06/15/99 - 04/30/00	Mechanical Oil Recovery in Ice Infested Waters (MORICE)-Phase IV	$64,000
10/01/98 - 09/30/99	Technology Coordinator	$100,000
12/01/99 - 11/30/00	Workshop Proceedings (CD-Rom): International Oil & Ice Workshop 2000	$25,000
02/01/00 - 01/31/01	Clean Boating Project	$10,000
05/01/00 - 02/28/01	Mechanical Oil Recovery in Ice Infested Waters (MORICE)-Phase V	$60,000
06/15/00 - 06/14/01	In Situ Determination & Monitoring of Hydrocarbons	$43,876
06/15/00 - 06/14/01	Computer Simulation of Spatial-Temporal Distribution of Dispersed and Nondispersed Oil Spills	$171,000
06/15/00 - 01/31/01	Nowcast/Forecast Program	$140,000
10/01/99 - 09/30/00	Technology Coordinator	$100,000
06/01/01 - 11/30/01	Response - Software Implementation	$60,633
06/01/01 - 11/30/01	Response - Software Implementation	$59,833
06/01/01 - 11/30/01	Response - Software Implementation	$60,633
03/15/01 - 10/31/01	Mechanical Oil Recovery in Ice Infested Waters (MORICE)-Phase VI	$80,000

Program Area	Institution Awarded	Modeling/Support of Modeling?
Tech	Jean Clarkin	
Tech	S.L. Ross Environmental	
Tech	PWS Science Center	
Tech	Cordova District Fishermen United	
Tech		
Tech	Tim Robertson	
Tech	UAF/SFOS/IARC	✓
Tech	Prince William Sound Science Center	✓
Tech	University of Miami/RSMAS	✓
Tech	Counterspill Research, Inc.	
Tech	SINTEF	
Tech	OSRI	
Tech	Alaska Clean Seas	
Tech	Cook Inlet Keeper	
Tech	SINTEF	
Tech	Arizona State University	
Tech	SINTEF	✓
Tech	University of Miami/RSMAS	✓
Tech	OSRI	
Tech	SEAPRO	
Tech	Alaska Chadux Corporation	
Tech	Cook Inlet Spill Prevention & Response, Inc.	
Tech	SINTEF	

continued

TABLE 6-1   Continued

Contract Term	Project Title[a]	Total[b]
10/01/00 - 09/30/01	Technology Coordinator[c]	$40,391
	Travel to Attend EPPR and Other Oil Response Meetings Abroad	$8,943
02/01/01 - 01/31/02	Nowcast/Forecast Program	$150,000
08/01/01 - 09/30/02	Ice Detection Technologies Using Radar	$100,000
09/01/01 - 09/30/02	Review of the OSRI Research Program	$100,000
TOTAL		$1,770,322

NOTE: All totals are approximate and are based on information provided by OSRI in February 2002.

[a]Descriptions of most projects can be found at the OSRI website <http://www.pwssc-osri.org>.

cost of $15,000/year to maintain. Information useful for local fishermen and boaters; part of dataset to be used in NC/FC model.

PWS Meteorological Data Collection—Development of a meteorological data collection capability within PWS. Partnership with several other entities. Input for NC/FC model.

Nowcast/Forecast Physical Modeling Project—This is a modeling project designed to implement and verify the Princeton Ocean Model (POM) for PWS. Long-term plans are to be able to model the physical environment so that the fate of oil can be predicted, therefore, that relevant biological resources at risk can be identified based on the projected fate of the oil and dispersants that might be used. This effort is discussed in depth in Chapter 7.

MORICE Phase 6.1—Coparticipant in the development and testing of an open-ice skimming machine (see Box 6-1).

Regional Atmospheric Model—Implementation, validation, and operation of a mesoscale atmospheric model for PWS. The primary use is to

Program Area	Institution Awarded	Modeling/Support of Modeling?
Tech/Eco/Edu	OSRI	
Other	PWS Science Center	
Tech	University of Miami/RSMAS	✓
Tech	Prince William Sound Science Center RCAC	
Other	National Academy of Sciences	

[b]This is the total spent on the project through early 2002; some projects continue.

[c]As of this entry, the technology coordinator is listed in all three programs (Ecology, Technology, and Education) rather than just in Technology, so Tables 5-1, 6-1, and 8-1 each include one-third of $121,175 in their total.

provide key atmospheric data to interface with NC/FC ocean/PWS circulation model.

Dispersion Impact Analysis—This activity uses the OSCAR model to conduct a wide range of dispersion scenarios to assist in future dispersant use planning and priority setting in real spill situations. The funds will be used by OSRI staff to run the model with various scenarios.

Oil and Ice Think Tank—A proposed meeting that will bring together leading experts to identify deficiencies in our current knowledge and capabilities for spill response in ice-infested waters.

Oil Response Inventory Project—A major effort, as indicated by the level of funding, was conducted by OSRI through the awarding of three contracts, for the period 6/1/01-11/30/01, to SEAPRO, to Alaska Chadux Corporation, and to Cook Inlet Spill Prevention & Response, Inc., for an inventory of oil spill response equipment. This included a software product designed to locate and determine the availability of such equipment in case of an oil spill.

BOX 6-1
ARCTIC OIL RECOVERY TECHNOLOGY:
THE MORICE PROJECT

Mechanical recovery of oil spilled in the presence of sea ice has always been difficult, and reliance on a strategy of burning spilled oil, when ice is present, has been the preferred alternative. When new ice is forming, oil is encapsulated into ice and is difficult to burn. Oil spilled under an ice sheet will be incorporated into the ice by natural freezing processes and can be neither burned nor mechanically removed. Therefore, such oil will be moved, along with moving ice, for eventual release in remote locations by natural melting processes. The presence of ice for most of the winter at Prudhoe Bay and even in Cook Inlet makes it difficult to provide for mechanical oil recovery there. Fortunately, the Valdez Arm has sea ice for only a brief few days in exceptional winters (once since the oil transport from Valdez began in 1977).

The MORICE program was a technology development initiative of industry, in which the principal producers of North Slope oil began to develop an apparatus for the mechanical recovery of oil spilled in ice-infested waters. The focus was to deal with a situation in which oil was floating on water that also had ice floes. The operational principles of skimmers for open water were adapted to the situation by designing an apparatus that would float and mechanically lift the ice floes on a conveyor belt, above the water level, and then would rinse the oil off of the ice floes, and the resulting oil-water mixture would be subjected to normal oil-water separation technology.

The device was conceived and a model was built and tested in the ice model tank of the Hamburgische Schiff Versuchs Anstalt (HSVA) in Hamburg, Germany. Performance was as predicted in model scale, and then a full-scale prototype was constructed and tested at Prudhoe Bay, Alaska. In this stage of the MORICE project, Phase IV, the participation of OSRI was initiated, along with the other industrial participants. Phase IV testing verified the operational principles, and set the limits for operation of the equipment. The equipment is not operational under conditions of freezing conditions, when frazil ice is being formed, and must be operated when ice is broken into floes of a size smaller than the width of the ramp apparatus that

---

RESPONSIVENESS TO MISSION

In general, the Applied Technology program is responsive to the OSRI mission. The committee had extended discussions about whether this part of the OSRI mission was properly focused and receiving adequate financial resources. The issue of balance in OSRI's portfolio of projects was

moves the ice floes above the waterline. The equipment can deal with a floe thickness of a certain maximum value, which unfortunately is much less than the ice thickness in Prudhoe Bay in late spring. Therefore, the applicability of the apparatus seems to be in areas such as Cook Inlet, but only when the ice floes are not driven by strong tidal currents. The conclusions of Phases IV and V of MORICE leave open the questions of where the technology may be applicable, and also confirm that many ice conditions are sufficiently severe and complicated as to frustrate mechanical cleanup of oil spilled in such conditions.

A second part of the evaluation of the MORICE apparatus was to be conducted by the prime contractor, SINTEF, in Svalbard. This field evaluation was a failure, due to the failure to anticipate the creep of the ice sheet under the combined effects of equipment loading and snow loading from an unexpected snowstorm.

The MORICE system appears to operate successfully under controlled, non-freezing air temperature conditions, within its limitations of ice floe size and thickness, and in the absence of ice movement or current flow. It may have applicability in the Great Lakes or parts of Canada, as well as in the European, Russian, or Caspian Sea regions. With reference to oil spills in Cook Inlet, where tidal currents are high, the apparatus has only the theoretical possibility of being useful, however. In Prudhoe Bay, the condition of non-freezing air temperatures along with the presence of broken sea ice does not exist except in summer, for some 60-75 days, which would represent the maximum window of opportunity for use of known methods of mechanical recovery of oil in ice.

OSRI participation in Phase VI of this program was ill-considered, and future participation of OSRI in the MORICE program is not recommended. Development of apparatus for recovery of oil on ice floes, or within ice, has at this time reached a plateau, with any future additional work awaiting a "break-through concept" that shows great promise. Costs of pushing toward a more effective apparatus would be very high, and any improvements on conventional technology would be incremental and marginal. Oil spill recovery equipment improvements should be viewed as beyond the scope of OSRI financial resources. Creative brainstorming of this intractable problem might, however, be worthwhile.

recognized by the Advisory Board and was partially dealt with by the previously described 40/40/20 targets for resource allocation. The monies expended in the applied technology area in recent years have been close to the established targets. However, as mentioned elsewhere, the classification of projects into these categories appears to be somewhat arbitrary and in some cases projects that might be classified in either category were assigned as needed to support the 40/40/20 formula. Overall since inception, nearly 40 percent of OSRI funds have been expended on projects categorized as applied technology. Trying to achieve advancements in the area of oil spill response in cold climates is a difficult undertaking that will require substantial investments of people, resources, and funds that far exceed the capabilities of OSRI. In apparent recognition of these limitations, OSRI often participated as a minor player in larger projects. The effectiveness and impact of these investments is limited.

Of the several OSRI projects and activities classified as applied technology, MORICE and Ice Think Tank were considered by the committee to be the most likely to have a direct impact on oil-spill-related issues in the Prince William Sound area. Although clearly focused on its mission, the MORICE project was marginal in effect and impact. Such large-scale technology development is very expensive and long-term, and the evaluation would have occurred without OSRI's contribution. The Ice Think Tank is a useful mechanism for defining the current state of the art and will provide OSRI with guidance for planning future directions for this portion of their program.

OSRI projects related to oil spill response that have a communications/education focus are effective. These projects provide an important public service and make a significant contribution to the long-term goal of reducing the release of pollutants to the environment. The program on small spill technology is a good example.

OIL SPILL RESPONSE—AN OSRI MISSION?

A key question is whether OSRI should be developing capabilities for real-time response to oil spills as a response organization or should it be playing a supporting role by providing research and advice to improve the efficiency of response and the range of tools available to responders. The OSRI mandate related to oil spill response has been subject to rather broad interpretation. In the committee's judgment, OSRI is not structured to be, nor is it provided with the authority, to act as a response organization. The committee concludes that OSRI should not operate as a real-time response mechanism and efforts expended in this arena will likely have little impact during an actual spill event.

Beyond the Nowcast/Forecast model (discussed in depth in Chapter 7), an example of an OSRI project that is primarily a real-time response activity is the purchase of database software for cataloguing and managing information related to spill equipment and training of personnel in various oil spill response organizations around the state. The cooperative spill response organizations for the three major areas where the oil spill response equipment is stockpiled (Prince William Sound, Cook Inlet, and Prudhoe Bay) normally would have such equipment lists and status reports; thus, it is not clear why OSRI was involved in this effort. In most cases, the regional U.S. Coast Guard office also maintains complete inventories of the spill response equipment capabilities in the region of their responsibility. This is a real-time response activity already served by other organizations and it is unclear how the OSRI purchase of a software package enhanced oil spill response capabilities.

FUTURE DIRECTIONS

The OSRI Applied Technology program should emphasize the development and improvement of techniques, material, and equipment that can be used to respond to and affect the cleanup of oil spills in cold marine environments.

Mitigation of Long-Term Effects

The people charged to respond to and clean up the *Exxon Valdez* oil spill (EVOS) faced myriad problems, and there are many opportunities for OSRI to evaluate cleanup methods and techniques suitable for the PWS and Alaska coastlines. During the oil spill, there was considerable controversy over the use of beach-cleaning agents, bioremediation products, etc. In some habitats, such as in mud-flat regions and in areas with heavy mussel beds, there was considerable debate and concern over the proper techniques to use.

Efficacy of Clean-Up Techniques

In addition to evaluating and recommending appropriate cleanup techniques to mitigate long-term impacts, OSRI could serve as an independent assessor of future cleanup techniques. Most cleanup activities will be handled by contractors, but there are opportunities to conduct appropriate studies to evaluate and assess the effectiveness of different cleanup techniques before spills happen. Alaska also has unique habitats and environmental conditions that are not favorable for standard oil spill response equipment. High tides and currents in excess of 1 knot create considerable difficulties for standard design booms and booming strategies.

The OSRI program should consider the evaluation of equipment and techniques that would be best suited for response under Alaska conditions.

Real-Time Assessment of Resources at Risk

OSRI should also continue to develop capabilities for rapid, real-time assessment of populations at risk. This real-time information could be used in support of resource agency input to the federal on-scene coordinator when priorities for oil spill response activities are being set.

Partnerships and Cooperation

In projects directly dealing with equipment and materials development, and in response strategies, OSRI should continue to seek cooperative input and participation with oil spill coops in Alaska. These are natural partnerships, and much of the real expertise in spill response is contained within the coops. OSRI should confine participation to programs where their special talents can make a real difference and not programs where OSRI becomes a subscriber. These special roles can best be identified by close collaboration with other programs and stakeholders.

In defining the roles that OSRI might play in oil spill response, it is important to understand that its role will be that of a supporting organization. Its knowledge and capabilities will be made available and mobilized through the unified command structure of the spill response. To more effectively make its capabilities available, it is important that OSRI develop relationships with the key spill response organizations, before a spill, to ensure that they are effectively included in the program if a spill occurs. This might be facilitated by the coop representatives, the NOAA science support coordinator, and the BP spill response coordinator on the OSRI Advisory Board.

Natural Resource Damage Assessment

Another component of the OSRI mission is to facilitate environmental assessment. The current OSRI project portfolio contains little in the way of projects that would assist in the conduct of natural resource damage assessment programs (see Chapter 2). One project that does fit into this category is the development of a PAH field monitor. OSRI should be identifying what key pieces of information are needed to conduct proper resource damage assessments, and then fund projects to address these needs including the improvement of assessment tools. For example, the waterfowl toxicity work was an effort to improve the use of biomarkers of exposure and could be considered a technology effort in support of future damage assessment work. Other suggestions are discussed in the report, *Oil in the Sea: Inputs, Fates, and Effects* (NRC, 2002), and could be developed through the strategic planning process.

7

Modeling

The central element of the present OSRI research and development effort is the Prince William Sound Nowcast/Forecast system (PWS-NC/FC). As shown in Table 5-1, the Predictive Ecology program provided about $533,000 to Nowcast/Forecast projects from FY98 to early FY02, out of about $1.6 million total program funds. Thus, 33 percent of predictive ecology funding went to support direct Nowcast/Forecast work. As shown in Table 6-1, the Applied Technology program provided about $763,000 to Nowcast/Forecast projects in the same period, out of about $1.8 million total program funds, or 43 percent. This may be an underestimate of the overall financial emphasis on Nowcast/Forecast: for example, it does not capture staff time (especially that of the technology coordinator, who is actively involved in running the model; funds spent since inception of this position in FY99 to early FY02 totaled $253,989, or an additional 14 percent of the applied technology budget). In addition, it does not reflect how other projects might be providing information that ultimately supports the model. Clearly, however, OSRI is devoting a significant proportion of its total efforts to modeling and its supporting components.

The PWS-NC/FC is a follow-on of a model system developed under the Sound Ecosystem Assessment (SEA) research program, which was funded by the Exxon Valdez Oil Spill Trustee Council (EVOSTC). Under the SEA program, a physical-biological model of PWS was developed with the goal of better understanding the physical-biological coupling and changes in the PWS ecosystem. This scientific understanding is needed to evaluate the potential impacts of oil spills on the PWS ecosystem, as well as other subarctic marine ecosystems. Thus, such a research and development effort is a fitting part of the mission of and was supported by OSRI.

As the SEA program was coming to a close, OSRI embarked on a new phase of model development. Whereas the focus of the SEA physical-biological model was to develop an understanding of processes and physical-biological coupling using hindcasts of specific periods, which were compared to field-collected data for verification, the present focus of the OSRI modeling program is in real-time nowcasting and forecasting on a daily basis. The BAA for this OSRI project (issued in early 1998) was brief, but requested the development of a "nowcast/forecast capability of currents and conditions that are relevant to the risk and costs of oil spills." The contract was awarded to the team that had developed the SEA model, which designed its proposal to include a real-time data acquisition and forecasting goal.

Central to the SEA physical-biological model system was a hydrodynamic model that predicts currents, temperature, and salinity distributions in four dimensions (3-dimensional in space and dynamic in time). The hydrodynamic model (Mooers and Wang, 1998) is an implementation of the Princeton Ocean Model (POM), widely used worldwide for such applications. The OSRI PWS-NC/FC POM application is undergoing continuing development and is being run in a near real-time nowcasting and forecasting mode at the University of Miami Rosenstiel School of Marine and Atmospheric Science (RSMAS, C. Mooers, PI).

OSRI recently added the existing Oil Spill Contingency and Response (OSCAR) model to the PWS-NC/FC. OSCAR was developed by SINTEF in Norway (Reed et al., 1995, 2000). OSRI purchased a license to OSCAR and houses a working copy of the model in the PWSSC, with the plan that OSRI staff and other investigators will run the OSCAR model in Cordova using data provided by the RSMAS POM through a web linkage.

OSRI has awarded a grant to the University of Alaska Anchorage (UAF, P. Olsson, PI) to develop a mesoscale atmospheric model that will produce a spatially and time-varying wind field to input to the hydrodynamic model at RSMAS. The existing Regional Atmospheric Modeling System (RAMS) model will be applied to PWS. The goal of the atmospheric modeling is to capture the orographic effects of the mountainous terrain of the islands and coastline around PWS, which is agreed to be the primary cause of the high variability in wind speed and direction across PWS at any given time. Describing this wind field variability in space and time is critical to accurate simulations of the hydrodynamics, as recognized by OSRI.

The conceptual design of the PWS-NC/FC is that the atmospheric model is to operate in real-time (a "nowcast/forecast") and pass wind fields to the hydrodynamic model. The hydrodynamic model will in turn prepare a Nowcast/Forecast simulation and pass both the wind and current predictions to the oil spill model. Both the atmospheric and hydrodynamic models will be automated to run every 6 hours, 24 hours a day. The oil spill model may be run at any time at OSRI to provide near real-time predictions of oil movements and fates. The running of the oil spill model is not automated, and the intention (for the near term) is that it be run by OSRI staff as needed.

The PWS-NC/FC is conceptualized to include biological and ecosystem model components, but these components are not yet developed. Whereas the SEA physical-biological model of PWS contained zooplankton and larval herring components, the principal investigators of those efforts are no longer involved in the OSRI effort. Thus, the biological modeling component is conceived as a future development effort. The present focus attempts to overlay the oil spill model's output on maps of fish species distributions developed using acoustical techniques.

MODEL PURPOSE AND ROLE IN OSRI MISSION

Given the extent of its financial commitment, OSRI clearly sees its modeling activities as critical elements of both the Applied Technology and Predictive Ecology programs. As discussed below, the committee has reservations about the purpose and perceived primary use of the modeling program as presently configured.

As currently envisioned, the models are planned for use in a predictive mode at all times as well as during a spill, and it is on this basis that they are classified as components of the Applied Technology program. For example, planned uses include providing real-time spill trajectories and 48-hour forecasts to guide oil-spill dispersant use decisions. That is, if the models demonstrate that the projected trajectory will take oil away from modeled concentrations of subsurface organisms, then it may be appropriate to use dispersants. Likewise, if the modeled subsurface dispersed oil concentrations are below some threshold value, then that information could also be used in making dispersant use decisions.

The problems with this thinking are two-fold. The first deals with the accuracy and validity of the model predictions and the fact that the models have not been developed to the point that they can be reliably used in this mode. At the present time, the existing modeling predictions are unreliable because they are being based on either a single wind record from a buoy located in the center of the Prince William Sound or 30-km-gridded atmospheric model forecasts of winds at the water surface. The surface current and oil trajectory is highly dependent on the accuracy of the wind field, and the model system lacks the necessary detailed wind information to support it. The value of the oil model output as a spill response tool is questionable also because it lacks a good validation of the hydrodynamic and wind models with measurements from different parts of PWS. OSRI and the principal investigators recognize these and other limitations, and plans are underway to develop more realistic wind forcing and increase the accuracy of model simulations. However, the ability to make accurate forecasts in real time is many years off, and the funding necessary to develop such a capability will be considerable.

One of the most important modules in the physical modeling program is proper validation. Based on the information provided, the field data collection programs do not appear to provide adequate real-world data to verify model predictions. If OSRI's long-term goal is to have their model used if future oil spills occur, they will have to show that the model accurately predicts the conditions that would occur within PWS.

An additional problem with the OSRI Nowcast/Forecast system deals with the acceptance of the model and OSRI's role in providing spill forecasts within the response community. Potential users in the response community include: the U.S. Coast Guard (USCG), the National Oceanic and Atmospheric Administration (NOAA), Alaska Department of Environmental Conservation, the responsible party, and other stakeholders involved in dispersant use decisions and other spill response issues (in situ burning, boom placement, skimmer deployment, etc). Legal authority for oil spill response offshore in U.S. waters has been given to the USCG, which relies on recommendations from NOAA for oil trajectory predictions when an oil spill occurs. NOAA utilizes its own modeling group and in-house models during all spill-response activities within U.S. coastal waters. Thus, this aspect of the OSRI Nowcast/Forecast system is redundant. However, there is an opportunity for OSRI to provide input to the response community by developing a better understanding of relevant processes, particularly of atmospheric and hydrodynamic transport and of the fates and effects processes involved.

In the crisis atmosphere following an oil spill, most participants recognize that a plurality of models would lead to different trajectory predictions, and these would add to the uncertainty and arguments about appropriate response measures. There is also an issue of liability if the OSRI model predicts that the oil will go in one direction with little harm or impact when the oil in fact goes in a different direction and causes unforeseen damages. The same problems of liability also apply to the NOAA model, but as an agency of the U.S. government, they are less likely to be a target of a lawsuit if their model predictions are inaccurate. Alyeska Pipeline Services Company (Alyeska) also has a similar model system (Alyeska Tactical Oil spill Model, ATOM), which it uses for contingency planning, drills, and spill impact evaluations, but it would defer to NOAA's modeling predictions in the event of a real spill response.

During a spill response, the initial deployment of equipment will be under the control of the Federal On-Scene Coordinator (FOSC) and the spill response coops. The strong probability is that they will depend on modeling input from NOAA's Seattle office, although NOAA may use information from other entities such as OSRI, as it is NOAA's job to integrate all available information. It is important to note that in real spill situations, there are always limitations on response equipment and personnel. Prioritization of response efforts is of primary importance. Unless strongly convinced otherwise, the FOSC is going to defer to NOAA's advice for setting the response priorities. OSRI would need to coordinate extensively with NOAA for its efforts to be useful in a spill response situation.

These same types of considerations and limitations relate to the use of the commercial spill trajectory program (OSCAR) purchased by OSRI. This can be a useful planning tool, and is also useful in developing and implementing oil spill response drills. In a real-world spill response, however, it would be questionable whether the USCG would defer to OSRI's predictions over those provided by NOAA through its science support coordinator, at least given the present status of the atmospheric and hydrodynamic models. In addition, NOAA's predictions are described as a probability distribution, reflecting the uncertainty in the wind and current forecasts input to the oil trajectory model. OSCAR's output is a single (deterministic) trajectory. Uncertainty in this forecast is not easily described without performing a sensitivity analysis of multiple runs manually implemented. As configured, it is doubtful that such uncertainty would be analyzed by OSRI and passed on to responders during a spill response in real time, as would be needed for OSRI to contribute constructively to a response.

At the present time, the NC/FC system is receiving real-time data from the data buoy in PWS, transmitting it to the University of Miami, where the nowcast/forecast hydrodynamic model is run, and then sending the data back to Cordova so that they can be coupled with the OSCAR model for making predictive trajectory runs. This is being done automatically every six hours. The value of having this information transferred back and forth on a daily basis is questionable, and maintaining and archiving data use funds that could be devoted to research. There has been a suggestion that this has been a temporary demonstration phase to show that they can transmit the data to Miami for processing and return on short notice. To continue this in perpetuity is of questionable value.

In summary, running the Nowcast/Forecast system is resource intensive from both a data and personnel viewpoint, and the OSRI legislative mandate does not justify this dedication of resources. Here are some relevant considerations to illustrate the issues related to real-time operational use:

- Who is responsible for the day-to-day operation of the ocean model now and in the future?
- How is this coordinated and adjudicated with other models that the government uses?
- What is the accepted, approved, and legally determined role of the results of this model?
- Who assumes liability for decisions based on the results of this model?
- Who will configure OSCAR on a real-time basis (type of oil, environmental parameters, etc.)?
- Who will run RAMS on a real-time basis?
- What real-time data networks provide input for the ocean, atmosphere, and fate models?
- Who will pay overtime for the operational infrastructure and maintenance?

Potential Uses of the OSRI Model System

Rather than focus the modeling in a real-time response mode, a much more appropriate focus for OSRI's modeling effort is in support of scientific purposes, such as contingency planning, training and community outreach, ecological risk assessment, understanding the Prince William Sound ecosystem, identifying data gaps, and designing experiments to obtain additional data on ecosystem function. Predictive forecasts are not needed for any of these purposes; rather, the model would be used in a hindcast mode, where wind and current conditions and oil spill behavior can be validated, thereby adding credibility to other model predictions (such as water column concentrations).

An example application is the planned dispersant impact analysis included in OSRI's current project list. However, while this is an appropriate research project for OSRI funding, it is not possible to evaluate the implications of changing (potentially increased) oil concentrations in the water column with application of dispersants without examining exposure and toxicity in more detail than is currently possible with OSCAR. The issue is the trade-off of floating and shoreline oiling of wildlife and habitats versus impacts on water column organisms. Chemical dispersants will increase concentrations in the water column. Whether this is significantly toxic is the question to resolve the dispersant use issue. OSRI plans to perform the dispersant scenarios in-house, without a competitive proposal process (G.L. Thomas, Oil Spill Recovery Institute. Personal communications, April 11, 2002.).

Because of the assumptions inherent in the construction of any model, an oil spill model is not an exact replica of the real world. The role of model building by OSRI should be viewed as advancing the state of models for oil spill response and for the prediction of ecological consequences. OSRI can make a contribution to oil spill response modeling by improving the knowledge base used to develop model algorithms. A second important use of OSRI models is to predict the ecological consequences of an oil spill, or of different possible oil spill scenarios. Model development should therefore include ecological effects.

STATUS OF NOWCAST/FORECAST MODEL COMPONENTS

Below is a summary of capabilities of the PWS NC/FC model components, the scientific validity and accuracy of the model simulations, and the degree to which the models represent or are pushing forward the state of the art.

Hydrodynamics Model

The hydrodynamic model in the SEA program (Mooers and Wang, 1998; Wang, 2001) was designed to be run as a hindcast for specific periods of time in order to evaluate physical-biological coupling and ecosystem dynamics. The model was forced with observational wind data from several stations, tidal data from three tidal stations, and freshwater runoff calculated from annual rainfall uniformly spread along the shoreline. The boundary was defined at Hinchinbrook Entrance (HE) and Montague Strait (MS) and forced with measured data at those locations. This implementation provided considerable understanding of the circulation dynamics within PWS.

The POM implementation under the OSRI program has been expanded to include the Alaskan Shelf from Yakatat Bay to Kennedy Entrance in Shelikof Strait. Earlier modeling and measurements at HE and MS showed inflow/outflow to be complex. GLOBEC investigations have identified eddies (20-200 km) that come up against the continental margin and onto the shelf, which have a role in exchanging waters in and out of PWS. There is large variability from year to year, and it is not always clear what is upstream or downstream on the shelf. On-shelf exchange between the coastal shelf current and PWS is complex and seasonally dependent.

As the nontidal circulation is driven by shelf processes, moving the boundary out to sea allows the longer-term circulation to be addressed.

This approach makes sense, as the only alternative is to force the model with high-resolution (vertically and temporally) measurement data at the entrances, and only one acoustic doppler current profiler is available at HE. However, intensive measurements could be made at the two entrances for specific events, which could be used to verify the dynamics within PWS. While efforts at validation have been made, there are still considerable differences and unknowns in the dynamics within PWS that should be addressed (such as circulation driven by freshwater inflow and local wind dynamics).

Only one wind station in central PWS is online in the NC/FC system at present (although other wind stations have existed and will be coming online), and any hydrodynamic model needs better forcing, particularly for winds. Winds vary tremendously across PWS because of orographic steering. There are important mesoscale features in the winds (gap winds, barrier jets) that will influence circulation and oil transport. The plan is to drive the hydrodynamic and oil models with gridded winds produced by a mesoscale atmospheric model.

Freshwater inflow also has an influence on circulation (Mooers and Wang, 1998), however including this forcing appears not to be in plans for immediate future for lack of sufficient data. A sensitivity analysis could be performed to determine how sensitive the model is to this forcing.

The recent hydrodynamic model validation (Bang et al., submitted) is comprehensive and is to be commended. Given the complexities in the underlying physical processes, periodic validation is necessary to calibrate the model and understand its strengths and limitations. It is not surprising that forcing functions have a dominant effect on the model predictions. One of the major weaknesses in the circulation model has been the modeling of through-flow from the Alaska Shelf. In particular, the assumption of equal and opposite volumetric transport at HE and MS is now known to be erroneous and will require closer attention in future modeling. Subsequent work has shown that there may be significant outflow through HE, and this needs to be incorporated in future modeling (Bang and Mooers, in press). Furthermore, (as noted above) the impact of coastal orography on the wind flow needs to be investigated and is likely to have a significant effect on any spill trajectory calculations. Finally, discrepancies in the predictions, mainly at the mesoscale and smaller scale compared with the basin scale, seem to reinforce the need for modeling the scale dependencies in the forcing functions using appropriate spatial models. The model will need further verification after the forcing is improved.

Atmospheric Model

The atmospheric model component has only recently been funded, and is not yet implemented in the NC/FC system. It will be an application of an existing mesoscale model, RAMS, which has been previously applied in other locations by the PI. The funded proposal and PI were favorably rated by both reviewers as technically sound, with the reservation that a postdoc to work on the project was not identified at the time of funding. This is an appropriate approach, as the wind forcing needs to be spatially variable at the scale being addressed by the mesoscale model, because of the high mountainous terrain surrounding PWS.

The entire domain will cover much of Alaska and the adjacent North Pacific, using a nested grid system of 64-km resolution in the outer domain, to 16-km resolution within it, and to 4-km in PWS. The nested design will invoke boundary issues, which will need to be resolved. In addition, the 4-km resolution within PWS may not be fine enough for the narrower passages between mountainous terrain and in passes. However, overall it is expected that the atmospheric forcing predicted by the RAMS will improve the hydrodynamics and oil spill simulations.

Oil Spill Trajectory and Fates Model

OSCAR (Reed, 2002) is a fairly comprehensive model for spill trajectory and fates calculations, accounting for most major physical and chemical processes and using algorithms that are state of the art and used in other similar oil spill fates models (e.g., ASCE, 1996; French et al., 1999). Fates processes included are advection, spreading, evaporation, emulsification, natural dispersion, dissolution, adsorption of soluble components and sedimentation, volatilization from the water column, degradation, stranding on shorelines, and such responses as booming, removal, and chemical dispersant application.

One important process, which is not included in OSCAR but should be, is the interaction of oil with suspended particulate matter and sedimentation of such material. This process has been shown to be important in a number of spills, and certainly could be important in much of Cook Inlet if modeling were expanded into that region, and in PWS at certain times of the year when stream, river, and glacier input are at their highest. Research is available for the development of an algorithm (Payne et al., 1987, 1989).

For the dispersant model, OSCAR uses an empirical dispersion rate based on limited field trials. Thus, there are many hypothetical functions in the algorithm, and much uncertainty in the results. Clearly more empirical data are needed to develop appropriate algorithms. Meanwhile, sensitivity analysis on the model would elucidate the uncertainty in present model results.

The focus of OSCAR is to model spill response scenarios and the fate of the oil resulting from such (Reed et al., 1995). The interface is designed to allow the user to play "war games" with response scenarios. Each simulation is set up individually, with databases of model input and a number of assumptions assigned by users. Thus, users need to be well-trained such that they can make appropriate selections of these uncertain inputs and interpret the results. Eventually, once the models are adequate, they should allow comparison to what is known from the *Exxon Valdez* spill. It will take some time (days to weeks) to run multiple scenarios with the present interface design, and these kinds of sensitivity analyses to varying model inputs will not be possible to run in a real-time response situation.

OSCAR does not account for the uncertainty in the velocity field and the resulting impact on the spill trajectory and travel times. A stochastic treatment of the velocity field or some kind of error propagation algorithm can potentially resolve this. The trajectory predictions will only be as good as the imposed velocity field and other environmental inputs. Given the sparse forcing data for the hydrodynamic model and wind field and spatial variations and uncertainties, it is impossible to describe such velocity distribution in deterministic detail. Clearly, a stochastic treatment is more appropriate as is routinely done in subsurface flow and transport in petroleum reservoirs and aquifers. The stochastic approach has also been applied to oil spill impact modeling (French et al., 1999; French McCay et al., 2002). Galt (1995) and Lehr et al. (2000) make the point that uncertainty must be taken into account in model simulations for oil spill response. The uncertainty quantification aspect is missing in the Nowcast/Forecast modeling exercise. This can lead to a false sense of security or overconfidence in modeling.

The OSCAR model seems to have been validated using a variety of synthetic problems by comparing the results with analytic solutions. Some limited validation of the trajectory and concentrations using field data has also been presented, although statistical analysis of goodness of fit is not performed.

OSCAR utilizes the POM current predictions for its simulations. Current data transfer from POM is done via an ftp (file transfer protocol) site, using an automated download onto an OSCAR-dedicated computer at OSRI. Presently, the wind data used is from a single record from a station in the center of PWS. The plan is to use the gridded predicted winds from the mesoscale atmospheric model, passed through the data transfer file from the POM.

The POM runs every six hours for a forecast of 48 hours. The current vectors are saved at simulation intervals of one hour on a 1-km grid (with 11 vertical layers in sigma coordinates) and downloaded to OSRI in Cordova every six hours. The download takes about 30 minutes and involves 20-30 megabytes of data per file. The data files are archived by date and time in a sequential set. The model automatically uses the latest hydrodynamic data available from the POM, stepping through applicable dated files sequentially.

Thus, OSCAR can be run using the POM input for as long as the POM forecast, which is now two days into the future. To run a hindcast, one needs to have the appropriate POM files for the period of interest. OSRI is saving a library of POM outputs for the present year, which might be used. The University of Miami group is preparing to make month-long files from the archives to use for hindcasts. However, there are no plans to run past years or an *Exxon Valdez* oil spill simulation period for potentially validating the model. Such an exercise would build confidence in the model system, as there is much information available on this oil spill.

Biological Effects Modeling

The exposure assessment model in OSCAR, as presently included in the NC/FC system, is highly simplified compared with the rigor involved in the transport and fates calculations. Exposure is evaluated as an index of oil thickness or concentration greater than a user-selected threshold, integrated and averaged within a defined polygon area. For surface-floating oil the index is m^2-hrs and for concentrations in the water, the index is ppb-hrs (presumably averaged over the water column). While mapping of such exposure indexes can show where resources are relatively more exposed and can be used to compare runs in a relative sense, it will be difficult to interpret such information in a response decision regarding using dispersants. The questions are (1) whether applying chemical dispersants will increase impacts on water column organisms, and (2) whether there will be a balancing decrease in impacts on wildlife and shoreline resources. The m^2-hrs of surface oiling needs to be translated into percentages of wildlife or numbers of birds oiled. The ppb-hrs of concentration exposure is highly subject to the volume and time over which the integration and averaging occurs. In addition, the ppb-hrs needs to be interpreted as an impact on water column organisms, which is a function of concentration of each chemical in the oil mixture, exposure time, and number of organisms exposed (i.e., a biological exposure model is needed). This kind of analysis will be needed by decision makers planning or during a response.

The original BAA for the biological effects component requested "technologies [that] demonstrate three dimensional trajectories as well as the resulting physical and biological environmental impact of dispersed and nondispersed oil in Arctic and subarctic marine environments." OSCAR does not yet have the biological impact component developed enough to meaningfully perform a comparison of the impacts of dispersed and nondispersed oil spills. At the moment the analysis would be limited to mapping of distributions of surface oil and subsurface concentrations in paired scenarios.

Resource mapping into an ArcView (Environmental Systems Research Institute software) geographic information system (GIS) (named RARE = Resources at Risk, which may be imported and overlaid on model results in OSCAR) using environmental sensitivity (ESI) maps and other information, is planned but not yet linked to the model. Meetings to discuss this plan were held in May 2002 in Trondheim, Norway. Also, consideration has apparently not been made of using the GRD (Graphical Resource Database) for PWS, which has been developed cooperatively by Alaskan and federal agencies and the Alyeska Pipeline Company over the last decade. The GRD is the intended coordinated resources-at-risk GIS database for PWS developed by the Regional Response Team (RRT). The GRD is used by Alyeska's oil spill model ATOM. OSRI should be working with the RRT, as the GRD could be read by the model system and is kept up to date with the latest information. Creating a new RARE database is a redundant exercise that could result in conflicting information. If OSRI wants to advise on spill response, they need to be more coordinated with the RRT.

In addition, the distributions of organisms that might be affected by a spill are highly variable in space and time. Static mapping of resource locations in a GIS map will be of limited use to an NRDA. More understanding of the coupled physical-biological system is needed for such purposes.

As part of the SEA program, there was a linkage between physical and biological modeling components. Plankton models (Eslinger et al., 2001; Jin et al., submitted) were run with the 3-D POM implementation by Mooers and Wang (1998). An understanding of the spring bloom temporal and spatial dynamics was gained that explained the focusing of herring populations in areas and times of plankton abundance. There were also bioenergetic models of pink salmon and herring by V. Patrick, D. Mason, and others. However, there has not been any incorporation of this or any other biological modeling into the current NC/FC modeling effort. It is unfortunate that the understanding gained in the SEA program will not be carried forward into the OSRI model system.

SCIENTIFIC SIGNIFICANCE AND IMPORTANCE

The models supported by OSRI are state of the art, but they are not unique models. The hydrodynamic and atmospheric models are applications of existing model code commonly used in other locations. The POM for PWS seems now to exist as two implementations: (1) the present OSRI model being operated at the University of Miami, and (2) the former SEA version of the model, now at the University of Alaska Fairbanks (Wang, 2001). The biological modeling component begun in SEA was lost, and a new effort will be required to bring this type of capability back to the OSRI system.

The OSCAR model is similar to the fates portion of the ATOM model, which has been in use by Alyeska Pipeline Company since 1990. ATOM is updated periodically and is equivalent to Applied Science Associates' SIMAP, a further development of the fates and biological effects portions of the NRDA model (French et al., 1996). In addition, NOAA HAZMAT has its own model GNOME applied in PWS; thus, there are redundant model systems in place now for PWS, at least in regards to oil fates. OSRI should rethink how its model system is to be used and whether it should in fact be a real-time nowcasting and forecasting system or a research tool. Also, while there are redundant oil fates models, there is a lack of hydrodynamic and atmospheric models, and these areas provide opportunities for OSRI to contribute.

FUTURE DIRECTIONS

Although the committee disagrees with the current focus of the OSRI modeling efforts on maintaining a real-time simulation capability, it sees good opportunities to shift the emphasis. In general, we recommend that OSRI focus on development of understanding and new algorithms that would improve the state of the art in oil spill modeling. Most of the algorithms used in oil spill models (generally) are based on or are similar to those developed more than 20 years ago by Mackay et al. (1980 and previous work). OSRI is in a unique position to fund algorithm development research, which is not generally funded by industry. The following are examples of topics that could be researched.

Hydrodynamics and Transport in Oil Spill Model

The POM is designed and set up to simulate the water circulation in subtidal waters, and does not handle the wetting and drying dynamics of intertidal flats. While such flats are limited in area in PWS, Cook Inlet and the Copper River Delta have vast flats. OSRI could make a contribution by funding development of modeling approaches to tidal wet-dry boundaries. Hydrodynamic models typically do not address the tidal wet-dry problem, and assume a shoreline of unchanging location.

The complexities of the COZOIL model (Reed et al., 1988, 1989) have not been included in OSCAR, likely in part because it requires considerable data on the shoreline characteristics. An algorithm of intermediate complexity to COZOIL and the simplification that is presently used in OSCAR is desirable to better estimate how much oil is retained by a shore segment and the dynamics as the tide moves in and out. Oil is known to combine with sediments and become incorporated in both intertidal and nearshore subtidal sediments. Understanding these dynamics is essential to predicting long-term impacts of oil spills on shoreline and nearshore biota.

Additional areas of research where OSRI funds could be used to elucidate and provide the basis of new and better oil fate algorithms include the influence of oil on wave height and transport by Langmuir circulation. Langmuir circulation was the subject of a recent NOAA workshop, because it is important to spreading and water-in-oil emulsification, as oil is concentrated in convergence lines.

Chemistry and Oil-Weathering Algorithms

Spreading

Spreading (which controls evaporation and entrainment) has been modeled for more than 20 years using algorithms that are only appropriate for calm water surface conditions. The algorithm is augmented by slick drift, oil droplet entrainment, and resurfacing, but significant data gaps exist in our understanding of oil-slick-spreading behavior.

Evaporation from (Thicker) Emulsified Oil

Evaporation from a thin (well-mixed) slick and the overall percentages of oil lost to evaporation can be predicted reasonably well. The rates of evaporation losses are often overpredicted where thicker oil slicks do not behave as well-mixed fluids. Usually this is because of water-in-oil emulsification (mousse formation). Under these conditions, evaporation is controlled by diffusion within the oil phase. This is an area that is currently not modeled at all in any oil-weathering model.

Emulsification

The formation of water-in-oil emulsions is only beginning to be understood at a fundamental level, and it cannot be predicted from first principles (Fingas et al., 1997). Most model predictions are based on empirical results from laboratory studies with a variety of oil types. Oil spill modeling would benefit from basic research focused on better understanding of these mechanisms, such that better algorithms can be developed.

Dispersed Oil Droplet–SPM Interactions

Dissolved component and suspended particulate material (SPM) interactions included in OSCAR are based on adsorption/desorption, but other research (Payne et al., 1987, 1989), has shown that whole oil droplet–SPM interactions overwhelm rates of dissolved component adsorption and sedimentation by several orders of magnitude.

Biological Effects

Modeling of biological effects has not been included in the NC/FC system. However, model development and oil spill effects research generally have identified several areas where greater mechanistic and quantitative understanding is needed in order to develop predictive models. Again, the following are examples of research areas.

An understanding of oil toxicity is critical to the prediction of biological effects. While acute effects on fish and invertebrates can be modeled with reasonable accuracy (French McCay, 2002b), long-term effects are being studied (Bodkin et al., 2002; Bue et al., 1998; Carls et al., 1999; Esler et al., 2002; Golet et al., 2002; Heinz et al., 1999; Marty et al., 1997; Roy et al., 1998; White et al., 1999) but have not yet been incorporated into modeling. There is a basic lack of understanding and quantitative information that would be needed to develop such models.

OSRI has focused on the distributions of specific species in PWS, with an aim of being able to predict resource exposure to oil. Research that develops a mechanistic understanding of biological distributions is highly desirable, as there is a basic lack of understanding of organism distributions and movements within PWS and elsewhere, such that they cannot be predicted. The ultimate goal would be to model the movements and exposure of organisms at several trophic levels. Presently, work is ongoing for several fish species, but other groups are not being addressed. The focus should be on those organisms most vulnerable: birds, marine mammals, and fish that utilize the nearsurface and nearshore (such as herring, pink salmon), ichthyoplankton and invertebrates in the surface waters, nearshore benthos, and intertidal organisms (both plants and animals).

An unanswered question that is often raised is: Do organisms avoid (or are they attracted to) oil? This would be a potential area for study, either using innovative laboratory studies or spills of opportunity. Such behavior could then be included in biological models and considered in NRDA.

Recovery of impacted species and success of restoration in augmenting impacted organisms are other areas that would be much in keeping with the legislative mandate and OSRI mission. The modeling of such recovery would be useful to NRDAs for future spills.

SUMMARY

The Nowcast/Forecast model is conceptualized primarily as a real-time spill response tool, and this and related efforts have been a large OSRI financial commitment and focus. The OSRI Advisory Board, with input from the OSRI leadership and stakeholders, will need to judge whether the emphasis on this modeling activity is an accurate reflection of its priorities and its interpretation of the OSRI mission. In the committee's view, the goal of OSRI's modeling activities should be to support research and, while OSRI may contribute needed information to responders, it should not compete in the response business. OSRI's modeling efforts should help advance the state of the art, including validation and analysis using hindcasting. Modeling can be used effectively to synthesize existing information and as a hypothesis-testing tool to define questions and identify needed research.

8

Education and Outreach Activities

One component of the Oil Spill Recovery Institute's Research and Development (OSRI R&D) program is public education and outreach, which focuses on improving public and professional understanding and knowledge of both technology and ecology R&D efforts. In 1997, the Advisory Board determined that public education and outreach was an appropriate activity to pursue under the OSRI mandate, and that a target of 20 percent of the annual R&D funding should be spent on these activities. In FY98-01, an average of 18.4 percent of the annual budget was spent on education and outreach. For FY02, a total of up to $315,000 has been allocated for education and outreach projects, which is very close to the 20 percent target. As most of these funds are awarded to "small" projects (i.e., dollar amount awarded is less than $25,000), the funds in the public education and outreach component of OSRI's program are awarded largely at the discretion of the OSRI director.

A variety of activities have fallen under public education and outreach, and in general, all appear to relate well to the OSRI mission. Projects included under the public education and outreach category, for years 1998-2002, are listed in Table 8-1. Further information on additional projects planned for 2002 is available in the FY02 OSRI Technology Coordinator's Report. Two of the larger project areas listed in that report, "Environmental Education K-12" (referred to as "Community Education" on the OSRI website) and "Science Planning Workshops," in fact encompass a variety of subprojects.

One notable ongoing activity within the environmental education K-12 category is "Science of the Sound," an environmental science enrich-

TABLE 8-1   Summary of OSRI-Funded Projects FY98-02 by Program
Area: Education and Outreach

Contract Term	Project Title[a]
01/30/98 - 03/15/98	Color Brochure
01/17/98 - 01/16/99	Web Page Maintenance
03/15/98 - 09/30/98	Newspaper Articles
06/01/98 - 07/31/98	Internship: Education Program
05/01/98 - 03/15/99	Video Production: "Sound Science"
07/01/98 - 06/30/99	Update of Museum Exhibit: "Darkened Waters"
10/01/98 - 09/30/99	Science of the Sound Education Program
02/01/99 - 01/31/00	Web Page Maintenance
05/01/99 - 09/30/00	Internship: Zooplankton Toxicity (Duesterloh)
05/10/99 - 08/30/99	Publication: Annual Report 1997-99
06/01/99 - 11/30/99	Internship: Waterfowl Toxicity Study
07/01/99 - 06/30/00	Fellowship Publications: 10-Year Anniversary of EVOS
08/01/99 - 07/31/00	Publications: Newsletters
01/01/00 - 12/31/00	Web Page Maintenance
07/01/00 - 06/30/01	Fellowship
04/15/00 - 12/01/01	Shorebird Educational Video: Celebrating Alaska's Shorebirds
04/14/00 - 04/13/01	Fellowship
07/01/00 - 06/30/01	Science of the Sound Education Program
07/01/00 - 06/30/01	Fellowship
07/01/00 - 08/30/00	Internship: Slides and Photo Organization
06/16/00 - 09/30/00	Internship: Network Administration
05/01/00 - 09/30/00	Publication: Annual Report 1999
09/15/00 - 04/30/01	Prince William Sound Meteorological Workshop
10/12/00 - 05/30/01	Workshop: Best Practices for Migratory Bird Care During Oil Spill Response
11/01/00 - 04/30/01	Publication: Annual Report 2000

Total Amount Awarded[b]	Program Area	Institution Awarded
$7,449	Edu	Outer Rim Publications
$9,000	Edu	PWS Science Center
$3,678	Edu	Outer Rim Publications
$883	Edu	PWS Science Center
$34,000	Edu	KAKM (Alaska Public Television)
$17,035	Edu	Pratt Museum
$25,000	Edu	PWS Science Center
$15,042	Edu	PWS Science Center
$3,306	Edu	PWS Science Center
$6,382	Edu	Jumping Mouse Publications
$6,000	Edu	University of California, Davis
$23,879	Edu	University of Alaska, SE
$3,276	Edu	In-house - OSRI staff cooperating with EVOS, PWS RCAC, Pratt Museum, Alaska Sea Life Center, and USFS
$5,184	Edu	Jumping Mouse Publications
$3,847	Edu	Phydeaux Multimedia, Monterey, California
$24,500	Edu	University of Alaska, Fairbanks/IAB
$10,000	Edu	KTOO TV
$25,000	Edu	University of North Carolina
$25,000	Edu	PWS Science Center
$25,000	Edu	University of Alaska, SE
$1,420	Edu	PWS Science Center
$8,671	Edu	PWS Science Center
$13,106	Edu	Jumping Mouse Publications
$38,474	Edu	Tim Robertson
$29,222	Edu	Production Plus
$12,000	Edu	Jumping Mouse Publications

continued

TABLE 8-1   Continued

Contract Term	Project Title[a]
11/01/00 - 08/01/00	Publication: Newsletters
	Fellowship
03/06/01 - 06/30/01	Workshop: Environmental Educators of Southcentral
03/06/01 - 10/31/01	Peer Listener Training Video
03/15/01 - 09/30/01	Public Policy: Phase 1
10/01/00 - 09/30/01	Technology Coordinator[c]
02/01/01 - 01/31/02	Science of the Sound
04/14/01 - 04/13/02	Fellowship
07/01/01 - 06/30/02	Fellowship
07/01/01 - 06/30/02	Orca Whale Rearticulation Project
08/01/01 - 12/31/01	Oil & Ice Think Tank
TOTAL	

NOTE: All totals are approximate and are based on information provided by OSRI in February 2002.

[a]Descriptions of most projects can be found at the OSRI website <http://www.pwssc-osri.org>.

ment program with a classroom component targeted at kindergarten to grade 6, and a summer camp component targeted at middle- and high-school students. Classroom activities are conducted with the elementary school in Cordova on a regular basis, and the program is taken to other PWS communities as feasible. Other activities that have been supported include contributions to the Pratt Museum in Homer for update of their "Darkened Waters" *Exxon Valdez* oil spill exhibit, to the native village of Eyak for the Orca whale rearticulation project, and production of several videos (including "Sound Science", and "Celebrating Alaska's Shore-birds"). Proposals for environmental education K-12/community educa-tion projects are submitted under an ongoing open BAA, which is posted on the OSRI website.

The Science Planning Workshop project is another of the larger com-ponents of public education and outreach, with up to $75,000 available for

Total Amount Awarded[b]	Program Area	Institution Awarded
$9,600	Edu	Jumping Mouse Publications
$24,500	Edu	University of Cambridge/Scott Polar Research Institute
$19,345	Edu	Alaska Natural Resource & Outdoor Education
$10,000	Edu	PWS Science Center RCAC
$0	Edu	Straight Arrow Consulting
$40,391	Tech/Eco/Edu	Walter Cox
$28,000	Edu	PWS Science Center
$25,000	Edu	University of North Carolina
$25,906	Edu	University of Alaska SE
$10,196	Edu	Native Village of Eyak
$12,400	Edu	D.F. Dickins, Ltd.
$581,692		

[b]This is the total spent on the project through early 2002; some projects continue.

[c]As of this entry, the technology coordinator is listed in all three programs (ecology, technology, and education), rather than just in technology, so Tables 5-1, 6-1, and 8-1 each include one-third of $121,175 in their totals.

FY02. Through this effort, OSRI has provided partial or full support for a number of science planning workshops related to oil spill technology, ecology, and education. Contributions to workshops or workshop proceedings have been funded under the Applied Technology program area, and in many cases these activities can be appropriately included in either area. Workshops funded under the public education and outreach component include the December 2000 Meteorological Workshop and the August 2001 Environmental Educator's Workshop on science and environmental programs in south-central Alaska. As with the Environmental Education K-12 project, there is an ongoing open BAA for solicitation of proposals for workshops. In addition to the workshop format, OSRI staff may want to consider offering occasional specialized short courses as an alternative to workshops, as this approach can be quite effective in conveying technical information to user groups.

A further public education and outreach activity is provision of support in the form of fellowships to graduate or postgraduate students, and in the form of internships to undergraduate and high-school students. Generally up to four awards have been given annually. The graduate fellowships may go to students on research projects that are also receiving OSRI funding, but they may also be awarded to students who are otherwise working independently of OSRI. The student internships have gone to support students undertaking a specific activity at OSRI/PWSSC. Overall, spending of OSRI funds on students appears to be well justified, providing that the research or internship projects relate to the OSRI mandate. Financial support for teachers from various backgrounds and educational levels, for attending workshops or doing projects related to oil spills, might also be a worthwhile use of funds.

Three additional projects that are included in the public education and outreach component of the OSRI program are (1) maintenance and updating of the OSRI website; (2) community and extension services, including publication and distribution of the OSRI newsletter and other OSRI literature, and public relations work as needed, and (3) production of the OSRI Annual Report. These projects have been contracted out in past years, and in FY02 a total of up to $40,000 has been allocated for continuing work. Although these activities are included under the area of public education and outreach for purposes of fund allocation, and certainly are necessary to the OSRI program, they could also be considered administrative functions. The staff of OSRI should exercise care when categorizing activities so placement of projects outside the administrative category does not constitute a mechanism to circumvent the 20 percent cap on administrative spending.

The contract to develop and maintain the OSRI website has generated some concern. Although there clearly was considerable effort initially put into the design of the OSRI website, more recent contract performance has not been satisfactory, as reflected in problems with the website, including a lack of up-to-date information. Staff of OSRI has identified this as a issue requiring attention, and they are attempting to address this in FY02.

FUTURE DIRECTIONS

The OSRI program for public education and outreach has included a relatively broad range of activities, and generally these activities are justified under the OSRI mandate and provide significant benefit at the scientific, educational or general community level. Primary and secondary school environmental education programs provide an excellent opportunity to enhance community understanding of the local ecosystem, effects of the *Exxon Valdez* oil spill, and in the longer term, threats from and responses to potential future spills. Fellowships and internships also are a well-justified use of funds. Support of workshops related to the OSRI mission is justified and OSRI might also consider sponsoring short courses on relevant topics and programs specifically for teachers as alternatives to the workshop format. The OSRI website reflects a great deal of initial effort, but is in need of attention to correct a number of deficits and bring it up to date; this is particularly important given the reliance of OSRI on its website to communicate solicitations.

9

Findings and Recommendations

When Congress established the Oil Spill Recovery Institute (OSRI), it did so in large part because the *Exxon Valdez* oil spill illustrated a serious need for improved information to deal with the special challenges posed by oil spills in cold-water environments. In its first five years, OSRI has produced some good results and it has had some problems. During the start-up phase of any new program some problems are to be expected, such as some unevenness in project quality and selection of some projects at the periphery of its mission.

Underlying the request for this outside review of OSRI is the question of whether the program should continue to exist past its legislated end in 2006. In the early years, once funding was provided, time was spent setting up policies and procedures. This learning curve is a part of the start-up of all research programs but this means that there is a relatively brief record for the committee to review (FY98 to FY01) and judge productivity and effectiveness. However, the advantage of conducting this review relatively early in the program's tenure is that it allows for some early course correction.

STRATEGIC PLANNING

Based on its review, the committee believes that OSRI has the potential to become a solid (albeit small) contributor to the quest for understanding of cold-water ecosystems and oil spills. There is excellent local support and involvement, and many unanswered questions could benefit from attention. To be effective, however, OSRI needs a new phase of strategic planning over the next decade, specifically to provide clear guidance about priorities and new areas of interest and to be sure that activities are closely aligned with its mission statement. To date, OSRI has focused heavily on its modeling component. Yet it could be doing much more to add to our understanding of oil and its effects on marine ecosystems, an area clearly within the OSRI mission but underserved by the current design. For instance, there are many areas that still need to be explored concerning the effects of shoreline cleaners and dispersants, bio-degradation, chronic effects on nearshore communities (flora and fauna), and long-term damage assessments.

PROGRAM BALANCE AND RESPONSIVENESS TO MISSION

The OSRI Advisory Board directs OSRI to have a 40/40/20 split among the applied technology, predictive ecology, and education/out-reach components of the OSRI program in an attempt to build balance into the program. This was a valid attempt to steer a new program, but it has not worked as intended. It has led to some arbitrary decisions about how portions of projects are split and recorded. For example, funding of projects that support the modeling work is accounted for in both the ap-plied technology and predictive ecology categories; yet the model clearly straddles the categories and does not fit neatly into either. It is not part of predictive ecology, because it does not yet have any linkages to deal with ecological or biological components of the system, although ultimately this is where the model might provide its greatest value. This illustrates that the categories are limiting; quality of work and relevance to mission should be the guiding criteria. The Advisory Board should revisit this allocation system. They should develop a long-term strategic plan that directs the program and assures that activities support the mission.

The predictive ecology and applied technology components are both generally responsive to the OSRI mission. Within each, there are a few examples that are less clearly directed to the mission, but overall, rel-evance is good. The efforts are fragmented because they are not linked by any themes or hypotheses that tie them to the mission. A strategic plan-ning effort led by the Advisory Board could provide concrete milestones to guide the programs.

Within the predictive ecology component, but not including the modeling activities, OSRI has funded a diverse set of projects, including studies and monitoring related to the effects of oil on waterfowl, herring and pollock, intertidal invertebrates on the Copper River Delta, rockfish, river otters, zooplankton/nekton, and other coastal resources. There remains great potential for research on long-term ecological effects of oil spills in Arctic and subarctic environments, particularly to improve understanding of and targeted monitoring of areas at risk of oil spills or chronic releases of petroleum. There is a great need for research to improve methods and strategies for bioremediation of oil-contaminated Arctic and subarctic marine and wetland ecosystems. There also is a need to better understand the physical, chemical, and biological fates (weathering) of petroleum in cold environments. Knowledge of biological resources, and how to protect them, is frequently lacking, as well.

The OSRI applied technology component also has supported a range of projects, focused on three goals: (1) development of tools to improve prevention and response to oil spills; (2) development of improved clean-up technologies; and (3) creation of models to assist with the deployment of equipment and personnel during an oil spill response for maximum effect and mitigation. Activities funded under the OSRI applied technology component include workshops, portions of the Nowcast/Forecast model, Mechanical Oil Recovery in Infested Ice Waters (MORICE), development of an in situ hydrocarbon monitor, a computer simulation of dispersant application during an oil spill, an inventory of oil response equipment, and a study of radar as an ice detection method. In general, applied technology is responsive to the OSRI mission.

Trying to achieve advancements in the area of oil spill response in cold climates is a difficult undertaking that will require substantial investments of people, resources, and funds that far exceed the capabilities of OSRI. In apparent recognition of these limitations, OSRI often participated as a minor player in larger projects. The effectiveness and impact of these investments is limited, however. For example, although OSRI helped support an evaluation of a skimmer designed to work in broken ice, such large-scale technology development is very expensive and long-term, and the evaluation would have occurred without OSRI's contribution. Within its technology emphasis, OSRI is better suited for broader explorations that ask, "What are the best approaches for preventing and mitigating oil impacts in Arctic and subarctic environments?" rather than trying to participate in the design of a specific technology. It might want to ask, "What are the regional broader impacts?" of any proposed activity. Even the smallest projects, like OSRI's small harbor project and many of the education/outreach projects, can have positive impacts at a broad, regional scale when approached with vision.

MODELING AND REAL-TIME OIL SPILL RESPONSE

The Nowcast/Forecast model is conceptualized primarily as a real-time spill response tool, and this and related efforts have been a large OSRI financial commitment and focus. In the committee's view, the goal of OSRI's modeling activities should be to support research and, while

OSRI may provide needed information to responders, it should not compete in the response business. To run a model every day when a spill might not occur for years is not an efficient use of resources. Instead, OSRI's modeling efforts should help advance the state of the art, including validation and analysis using hindcasting. Modeling can be used effectively to synthesize existing information and as a hypothesis-testing tool to define questions and identify needed research. Instead of being another oil trajectory model, it could be recast into a research model that helps people understand the system, its functions, and forcings, and think about "what if." It would become useful to researchers, risk assessors, and planners, as well as providing input to responders.

The committee wants to stress that despite its concerns about the degree of emphasis given to model development and its operational focus, the work done to date has been good. But other models already exist that can and will be used to help make judgments about oil spill trajectories in the event of a spill. Much work would still be needed to demonstrate that NC/FC is truly better than other existing and developing models. It lacks essential linkages with biological aspects of the system, which are critical if it is to be used from a resource protection perspective. Given OSRI resources and the expense of model development, continuing with model development as the primary focus will keep OSRI from expanding its other activities. A more diverse program is needed.

Modeling capability can be an important element of oil spill preparedness and response, and there are questions and problems within this realm that OSRI could help address. For example, one good use that could be developed from the current base would be modeling of long-term biological effects and physical-biological couplings. As another example, an understanding of hydrodynamics is needed to understand and predict change in the ecological system, and atmospheric forcing is critical to understanding the circulation of Prince William Sound and oil transport within it.

GEOGRAPHIC FOCUS

OSRI had done limited work in true Arctic environments. While the committee is fully aware that there are many needs and unanswered questions related to oil in truly Arctic marine settings, we understand OSRI's decision to focus on Prince William Sound, Cook Inlet, and the northern Gulf of Alaska. OSRI is a small program and must make choices about allocation of resources to achieve some critical mass of work that has an impact. In all its work, but especially in Prince William Sound, there will be a need for continued coordination with other research programs (e.g., the *Exxon Valdez* Oil Spill Trustee Council's Gulf Ecosystem Monitoring program and the new North Pacific Research Board).

PROGRAM OVERSIGHT

The committee has reviewed the minutes of most Advisory Board meetings, to get a sense of its engagement with OSRI operations. The minutes show the maturation of the Advisory Board and OSRI, as it moved from its early planning phase to now. The Advisory Board has had frank discussions about a number of important issues and problems and has shown a willingness to make changes and institute new procedures when necessary. The OSRI Advisory Board is composed of representatives of agencies, each with its own mission and sometimes differing needs. Because not all members of the Advisory Board have detailed scientific experience, the creation of the Scientific and Technical Committee was a sound step, so there is a small group capable of providing in-depth scientific insight and leadership. The STC is an important part of the checks and balances of the OSRI process, although the committee was less able to evaluate this body because of limited time and fewer written records. The STC should continue to play a key role in quality assurance of the program. It should have an active role in judging the quality and appropriateness of medium and large proposals. Term limits and clear procedures for selecting new members should be implemented, to help ensure that the STC remains an independent voice in the OSRI program.

FAIRNESS ISSUES

Based on its meeting minutes, the Advisory Board is already aware that there is a perception in the science community of possible conflict of interest and fairness issues within OSRI. Even if this is just an image problem, based on outdated or misinformation, the negative perceptions affect the program. They diminish the program's appeal to qualified outside scientists and cast a shadow over its credibility.

Part of dealing with the negative perceptions about OSRI will necessitate dealing with the close relationship between OSRI and the Prince William Sound Science Center (PWSSC). These two organizations are clearly linked, and this is not necessarily inappropriate. Because OSRI and the PWSSC are both small organizations, located in a small and somewhat isolated community, there are real cost efficiencies to be gained by sharing staff and facilities. However, because OSRI grants significant funding to the PWSSC and because the two organizations share staff, including the director, this feeds a perception of potential impropriety and can set up opportunities for real problems. Negative perceptions will take time to overcome, but attitudes can be changed by open communications and by following all required procedures carefully.

ACTIONS NEEDED

Although OSRI is a small program within the larger scientific context and it has had some problems that need to be addressed, OSRI is doing some good work and it is a big influence in Cordova, Alaska. Some of its educational programs deserve special credit for building a strong community partnership.

What follows are the committee's main findings and recommendations, drawn from each chapter. The focus is on the overall OSRI portfolio and procedures, and not specific projects, with the exception of a more detailed exploration of the modeling efforts because these are such a large component of the OSRI program. The items vary in their specificity; however, all are intended to help the Advisory Board and OSRI staff strengthen the program as it evolves, especially if it is decided that the program should continue past its currently scheduled sunset in 2006.

ORGANIZATION AND ADMINISTRATION

Finding 1: Overall, the organization and administrative structure of the Oil Spill Recovery Institute are appropriate for the mission. As with any new program, there were start-up problems, and some problems remain (addressed in the following findings and recommendations). But in general the Advisory Board has discussed problems frankly and implemented procedures to address them (e.g., the Grants Policy Manual).

Recommendation: The committee recommends continued strong oversight from the Advisory Board and the Scientific and Technical Committee. These perspectives and added expertise are important to ensuring that OSRI fulfills its mission and generates high quality research that is useful over time in improving our understanding of, and ability to respond to, oil spills in cold marine environments.

GRANT AWARD POLICIES AND PROCEDURES

Finding 2: Solicitations for proposals, especially some of the Broad Area Announcements (BAAs), do not always accurately and adequately describe the scope of the research being requested. Confusion in this area in the past may have led to some of the frustration expressed by investigators who felt unfairly treated in the OSRI grant review process.

Recommendation: The success of the OSRI research program requires careful attention to the writing and administration of the proposal process. BAAs are appropriate for some objectives of OSRI (e.g., education programs, fellowships and internships, and in cases where novel solutions are being solicited). However, RFPs that are specific in objective are likely to be more effective for many of the projects funded by the OSRI. Well-written RFPs would provide potential bidders with adequate information to respond to long-term strategic goals.

Finding 3: The solicitation process has not been effective in reaching a broad audience, which is critical to ensure high-quality science and technology. BAAs and RFPs were not well advertised or released in a regular cycle. Whether true or not, there is a perception in the scientific community that some people have advance knowledge of upcoming solicitations.

Recommendation: Improvements in the process used to solicit research and technology development proposals are needed.

- Solicitations for proposals should be released on a standard schedule and should be open for a reasonable amount of time (e.g., three months). Additional effort is needed to ensure BAAs and RFPs are independently developed and in concert with the mission and objectives of OSRI.
- Solicitations should be advertised nationally, which will help garner wider responses and increase awareness of OSRI goals and objectives.
- Implementation of existing procedures should be strict, to assure potential bidders that the procurements are not "wired."
- OSRI should increase its use of the RFP rather than BAA process when specific research is desired.
- A schedule for review and decision making should be established so that investigators who have expended considerable time and money will get a timely response to their submittal. Unsuccessful bidders should be notified of the final selection and provided some feedback regarding the reason for their unsuccessful bid.
- OSRI should define a minimum distribution list of organizations, publications, and individuals for the dissemination of

BAAs and RFPs. These should include written and electronic media, both within Alaska and across the United States. The Arctic Research Consortium of the United States (ARCUS), based in Fairbanks, Alaska, manages a listserve that reaches more than 3,000 people and is one example of a mechanism that can be used. Consideration should be given to include distribution to other countries where expertise resides.

Finding 4: As of August 2002, the list of OSRI publications includes about 44 items, either published or in press, of various types including journal articles, conference proceedings, abstracts, videos, and assorted maps and guides. The number of articles in refereed journals appears to be increasing. Thus OSRI research is starting to make its way into the refereed literature.

Recommendation: OSRI leadership should continue to place high emphasis on funding research likely to be published in peer-reviewed publications that have broad distribution. The nature and number of publications should be monitored into the future, using publication rates as one measure of the quality of the program.

Finding 5: The 20 percent limit on overhead discourages some potential researchers from applying for funds. For example, many applicants at academic institutions would be restricted from applying given that constraint.

Recommendation: The Advisory Board has discussed this potential problem and agreed that additional overhead may be paid at the discretion of the OSRI director. This policy should be made clear in future advertising of BAAs and RFPs. For example, a statement could be added that government-audited indirect cost rates at academic institutions will be honored if a proposal is selected and funds are available.

PROGRAM PLANNING

Finding 6: OSRI's most recent strategic planning document was issued in 1995. Since the 1995 document, there has been an evolution in interpretation of the OSRI mission and associated goals and the 1995 strategic planning document no longer directs OSRI's future. Annual plans and the Technical Coordinator Report comprise an important part of the organization's planning process, but serve more as records of past and current activities rather than to help set a course for the future.

Recommendation: A revised five-year strategic plan should be developed, led by the Advisory Board, and perhaps for efficiency assigned to a subgroup such as the work plan committee, with strong input from OSRI staff and the Scientific and Technical Committee. The planning process should explicitly consider the issue of "top-down" and "bottom-up" approaches and provide clear guidance on how the Advisory Board wishes the administration to proceed. Annual plans should be continued and expanded to provide an update on progress toward long-term strategic goals as well as a summary of projects. Together, these documents will not only guide the program but will be very useful to Congress when it decides on the future of the program.

The strategic plan should

- contain the history and evolution of the OSRI mission and objectives including the rationale for these changes in course.
- evaluate the 1995 recommendations/priorities versus progress to date or explicitly declare them no longer applicable.
- provide a clear overarching vision for OSRI for the rest of its legislative life, including an explanation of how funded activities fit into this vision, and describe future challenges that OSRI might address should it continue.
- explicitly identify mission-related goals and the activities to be pursued to reach these goals.
- provide a detailed time line and milestones by which progress toward the goals can be assessed on regular basis.

PREDICTIVE ECOLOGY

Finding 7: OSRI predictive ecology projects are generally responsive to the OSRI mission. There needs to be more synergism and connectedness between the field studies and the predictive ecology modeling.

Recommendation: The predictive ecology portfolio should be continued with improvements in design and in scope as detailed elsewhere in this report. Outputs from observational field studies should be closely aligned with the modeling efforts to ensure that relevant parameters are being measured and that data is collected on spatial and temporal scales consistent with integration into the models. OSRI should develop a niche for its studies in larger regional ecosystem studies, where it can provide leverage and synergy.

Finding 8: The Copper River Delta study, Prince William Sound resource monitoring, and the sensitivity mapping have produced quality results that increase our understanding of ecosystems in the area. The work is well designed and is providing valuable information about the ecology of forage species of invertebrates that are preyed upon by migratory birds.

Recommendation: All ecosystem monitoring should strive to develop a better understanding of ecosystem functioning and structure. Such studies require rigorous study design. Inventories are of limited value as a first order assessment of resources present, since the biological patterns vary over time and space. To enhance the usefulness of the Copper River study, chemical analyses (hydrocarbons, metals, and nutrients) should be done to allow comparison if there is a spill upriver or in the Gulf of Alaska.

Finding 9: OSRI's biological monitoring generally covers the four dominant biomasses in Prince William Sound and the Copper River Delta (Pacific herring, pollock, *Neocalanus copepod*, and *Macoma balthica*). While this approach provides insights about financially important species, it does not provide a comprehensive picture of the ecosystem. Many research programs consider monitoring to be within their missions, so coordination with monitoring efforts by other organizations is particularly important.

Recommendation: When OSRI supports biological monitoring programs in Prince William Sound, it should cover a broader spectrum of the ecosystem. For example, zooplankton data need to be integrated with synoptic nutrients and phytoplankton data. Herring and pollock studies should evaluate predator and prey distributions and abundances, so that the distribution and ecological data for the two species can be integrated into the ecological and food chain models for the sound. Results should be integrated with other fisheries and ecological studies in the sound, including the *Exxon Valdez* Trustee Council's GEM program, projects of the newly evolving North Pacific Research Board, and other ecosystem-oriented research programs. Where possible, monitoring should be expanded to include a wider geographic area of the sound and seasonal observations.

APPLIED TECHNOLOGY

Finding 10: The projects funded within the Applied Technology program are generally responsive to the OSRI mission. OSRI purchases of existing software packages have limited impact and only marginally contribute to the OSRI mission.

Recommendation: OSRI Applied Technology projects should concentrate on the development or improvement of techniques, materials and equipment that affect the efficacy of the oil spill prevention and clean up response. A component of the overall OSRI portfolio should support natural resource damage assessment, such as efforts to improve assessment tools. Purchases of existing software or other off-the-shelf items should be carefully considered to be certain they are relevant to the OSRI mission and add value to the program.

Finding 11: OSRI is not structured, funded, or mandated to be a real-time oil spill response organization.

Recommendation: OSRI should not fund activities, projects, or purchases that have a primary real-time response justification. OSRI should be encouraged to develop a close relationship with oil spill response organizations to clearly define an appropriate and realistic role in the case of an actual spill.

Finding 12: The education and outreach activities supported by OSRI are generally well designed and effective. There are real opportunities available to create educational activities with an applied technology orientation.

Recommendation: The impact of the Applied Technology program could be enhanced by creating additional education and outreach activities, such as short courses, and by designing new activities with a technology focus.

PROGRAM BALANCE

Finding 13: Although creation of the 40/40/20 split among predictive ecology, applied technology, and education/outreach was an appropriate initial attempt to build balance into the overall OSRI program, it is not accomplishing its intended effect. In some cases, project classification based on the 40/40/20 criteria appears to be arbitrary and the requirement is actually undermining the goal of project balance.

Recommendation: Changes should be made in the 40/40/20 classification requirement. Either projects should be classified objectively as to category or the targets for allocation rescinded. Award selection should be based on quality and relevance to mission. Balance should be assessed year to year but overall balance should be judged based on the entirety of the program's portfolio over the lifetime of the Institute.

MODELING

Finding 14: The OSRI-supported Nowcast/Forecast (NC/FC) model system is conceptualized primarily as a real-time oil spill response tool.

Recommendation: The mission of OSRI should be to fund research, and while OSRI may contribute needed information to responders, it should not compete in the response business. OSRI's modeling efforts should work to advance the state of the art, including validation and analysis using hindcasting. Modeling can be effectively used to synthesize existing information and as a hypothesis-testing tool to define questions and needed research. It is more appropriate that OSRI use modeling in research, contingency planning, and ecological risk.

Finding 15: The NC/FC system is not a validated and accepted response tool.

Recommendation: If OSRI intends to continue with this role for the model system, it will need to be validated and the potential user community will need to be persuaded that it is a worthwhile tool.

Finding 16: Considerable OSRI funding has been committed to implementation of existing modeling technology and knowledge, such as model algorithms that have been incorporated in other modeling systems applied elsewhere as well as in Prince William Sound.

Recommendation: The incorporation of existing models is appropriate if it is clearly related to research goals. However, OSRI would make a greater contribution by funding research that advances the state of the art of modeling—for example, by improving algorithms for oil transport, fates, and effects—rather than funding development of code to implement (or reimplement) existing capabilities and knowledge. OSRI could play a role in basic research of modeling, which is typically not funded by industry.

Finding 17: Model validation efforts have been limited to date, in part because of the lack of data to input to the models. These limitations have been recognized by OSRI and attempts are being made to improve data inputs to the model system, such as the addition of the atmospheric modeling component to provide spatially varying wind fields to the hydrodynamic and oil spill models.

Recommendation: OSRI should make it a priority to validate the entire system together, as well as each model alone. In addition, the comparison of model predictions and observations should be described statistically, where goodness of fit is measured and uncertainty is presented quantitatively.

Finding 18: The present emphasis on real-time forecasts every six hours is inappropriate and unnecessary, as there is little understanding gained from each simulation. In addition, the model system as it stands is deterministic—that is, it produces single simulations with no measure of uncertainty.

Recommendation: OSRI's modeling efforts would be more productive and cost effective if they focused on intensive observational periods (of a month or so duration) to develop a better understanding of the system by comparing model results and observations. In addition, the model system needs to be applied in a probabilistic/stochastic mode, with quantitative uncertainty estimation. This is appropriate whether the application be spill response or ecological risk assessment.

Finding 19: There has been little effort devoted to incorporating biological effects into the modeling or in quantifying recovery aspects of Prince William Sound after the *Exxon Valdez* oil spill. Early OSRI studies did evaluate physical-biological coupling, but this work has not been continued.

Recommendation: OSRI should plan for and implement biological modeling into the program.

Finding 20: The current focus on the atmospheric model development is appropriate and needed for better understanding of hydrodynamics and oil transport, as well as the implications to biological communities.

Recommendation: As OSRI's modeling activities evolve, continued work will be needed to improve the atmospheric components and continued attention to freshwater runoff will be needed for the hydrodynamics to be understood and predicted.

Finding 21: There has been little coordination between models until a recent meeting of principle investigators in April 2002. OSRI is writing a strategic plan for modeling based on the workshop.

Recommendation: There is a need for more and better planning of the model system, in writing, and with specific goals, deliverables, and a timeline. For example, the coupling of the hydrodynamics and oil transport needs to be tight, to be sure no redundancy of forcing (wind drift, Ekman flow) is included.

Finding 22: Coordination with other programs and organizations involved in modeling has been limited.

Recommendation: There should be increased coordination with the modeling components of the Gulf Ecosystem Monitoring (GEM) program, GLOBEC, SALMON and other research programs, especially regarding boundary conditions and coordination with larger scale atmospheric models, and better coordination with NOAA and Alyeska.

EDUCATION AND OUTREACH

Finding 23: Activities included under the Public Education and Outreach program of OSRI cover a broad range and they are a valuable contribution to the overall OSRI program.

Recommendation: Education and outreach activities should be continued, similar to current efforts. Staff must ensure that these activities support the OSRI mission, meaning that they must have clear links to oil spills, their prevention and response, and environmental impacts and protection. There may be special opportunities for more technology-focused activities and activities directed at teachers.

Finding 24: Although most of the projects included in the Education and Outreach program generally fall within the OSRI mission, some may be more appropriately considered administrative functions (e.g., annual report development, computer technology, and website maintenance).

Recommendation   OSRI management, under the direction of the Advisory Board, should frankly discuss the limitations imposed by the 20 percent overhead cap and give explicit approval that certain overhead-type activities can or cannot be conducted under the auspices of the Education and Outreach program.

References

ASCE Task Committee on Modeling Oil Spills. 1996. State-of-the-art review of modeling transport and fate of oil spills, Water Resources Engineering Division, ASCE. *Journal of Hydraulic Engineering* 122(11):594-609.

Atlas, R.M. 1995. Petroleum biodegradation and oil spill bioremediation. *Marine Pollution Bulletin* 31:178-182.

Bang, I., and C.N.K. Mooers. In press. The influence of several factors controlling the interactions between Prince William Sound, Alaska, and the Northern Gulf of Alaska. *Journal of Physical Oceanography*.

Bang, I., S. Vaughan, and C.N.K. Mooers. Submitted. Initial steps towards validation of a seasonal cycle simulation for Prince William Sound circulation (flow and mass) fields. Submitted to *Continental Shelf Research* (14 Jan 2002).

Bodkin J.L., B.E. Ballachey, T.A. Dean, A.K. Fukuyama, S.C. Jewett, L. McDonald, D.H. Monson, C.E. O'Clair, and G.R. VanBlaricom. 2002. Sea otter population status and the process of recovery from the 1989 *Exxon Valdez* oil spill. *Marine Ecology Progress Series* 241:237-253.

Bragg, J.R., R.C. Prince, E.J. Harner, and R.M. Atlas. 1994. Effectiveness of bioremediation for the *Exxon Valdez* oil spill. *Nature* 3689:413-418.

Bue, B.G., S. Sharr, and J.E. Seeb. 1998. Evidence of damage to pink salmon populations inhabiting Prince William Sound, Alaska, two generations after the *Exxon Valdez* oil spill. *Transactions of the American Fisheries Society* 127:35-43.

Carls, M.G., S.D. Rice, and J.E. Hose. 1999. Sensitivity of fish embryos to weathered crude oil. Part I. Low-level exposure during incubation causes malformations, genetic damage, and mortality in larval pacific herring (*Clupea pallasi*). *Environmental Toxicology and Chemistry* 18(3):481-493.

Cooney, R.T., K.O. Coyle, E. Stockmar, and C. Stark. 2001. Seasonality in surface-layer net zooplankton communities in Prince William Sound, Alaska. *Fisheries Oceanography* 10(Supp.):97-109.

Dickens, D.F. 2002. Development of a Draft Research Agenda for Oil Spill Response in Ice-Covered Waters. Final Technical Report. Oil Spill Recovery Institute, Cordova, Alaska.

Esler, D., T.D. Bowman, K.A. Trust, B.E. Ballachey, T.A. Dean, S.C. Jewett, and C.E. Charles O'Clair. 2002. Harlequin duck population recovery following the *Exxon Valdez* oil spill: progress, process and constraints. *Marine Ecology Progress Series* 241:271-286.

Eslinger, D.L., R.T. Cooney, C.P. McRoy, A. Ward, T.C. Kline, E.P. Simpson, J. Wang, and J. Allen. 2001. Plankton dynamics: observed and modeled responses to physical conditions in Prince William Sound, Alaska. *Fisheries Oceanography* 10(1):81-96.

Fingas, M., B. Fieldhouse, and J.V. Mullin. 1997. Studies of water-in-oil emulsions: stability studies. Pp. 21-42 in *Proceedings of 20th Arctic and Marine Oil Spill Program (AMOP) Technical Seminar*. Environment Canada, Ottawa, Ontario.

Foght, J.M., and D.W.S. Westlake. 1982. Effect of the dispersant Corexit 9527 on the microbial degradation of Prudhoe Bay oil. *Canadian Journal of Microbiology* 28:117-122.

French, D., M. Reed, K. Jayko, S. Feng, H. Rines, S. Pavignano, T. Isaji, S. Puckett, A. Keller, F. W. French III, D. Gifford, J. McCue, G. Brown, E. MacDonald, J. Quirk, S. Natzke, R. Bishop, M. Welsh, M. Phillips, and B.S. Ingram. 1996. The CERCLA type A natural resource damage assessment model for coastal and marine environments (NRDAM/ CME), Technical Documentation, Vol. I - Model Description. Final Report, submitted to the Office of Environmental Policy and Compliance, U.S. Dept. of the Interior, Washington, DC, April, 1996, Contract No. 14-0001-91-C-11.

French, D., H. Schuttenberg, T. Isaji, 1999. Probabilities of oil exceeding thresholds of concern: examples from an evaluation for Florida Power and Light. pp. 243-270 In: Proceedings: AMOP 99 Technical Seminar, June 2-4, 1999, Calgary, Alberta, Canada, Emergencies Science Division, Environment Canada, Ottawa, ON, Canada.

French McCay, D. 2002a. Modeling evaluation of water concentrations and impacts resulting from oil spills with and without the application of dispersants. International Marine Environmental Seminar 2001. *Journal of Marine Systems*, Special Issue 2002.

French McCay, D.P. 2002b. Development and application of an oil toxicity and exposure model, OilToxEx. *Environmental Toxicology and Chemistry* 21(10):2080-2094.

French McCay, D., N. Whittier, S. Sankaranarayanan, J. Jennings, and D. S. Etkin, 2002. Modeling Fates and Impacts for Bio-Economic Analysis of Hypothetical Oil Spill Scenarios in San Francisco Bay. Proceedings of the Twenty Fifth Arctic and Marine Oil Spill Program (AMOP) Technical Seminar, Environment Canada, Calgary, AB, Canada, 2002.

Galt, J.A. 1995. The integration of trajectory models and analysis into spill response systems: the need for a standard. Pp. 499-507 in *Proceedings of the Second International Oil Spill Research and Development Forum*. International Maritime Organization, United Kingdom.

Golet, G.H., P.E. Seiser, A.D. McGuire, D.D. Roby, J.B. Fischer, K.J. Kuletz, D.B. Irons, T.A. Dean, S.G. Jewett, and S.H. Newman. 2002. Long-term direct and indirect effects of the *Exxon Valdez* oil spill on pigeon guillemots in Prince William Sound, Alaska. *Marine Ecology Progress Series* 241:287-304.

Heintz, R.A., J.W. Short, and S.D. Rice. 1999. Sensitivity of fish embryos to weathered crude oil. Part II. Increased mortality of pink salmon (*Oncorhynchus gorbusche*) embryos incubating downstream from weathered *Exxon Valdez* crude oil. *Environmental Toxicology and Chemistry* 18(3):494-503.

Jin, M, J. Wang, P. Simpson, A.E. Ward, P. McRoy, and G. Thomas. Submitted. A 3-D coupled physical-biological model and its application to the spring plankton bloom of 1996 in Prince William Sound, Alaska. *Journal of Geophysical Research*.

Lehr, W.J., D. Wesley, D. Simecek-Beatty, R. Jones, G. Kachook, and J. Lankford. 2000. Algorithm and interface modifications of the NOAA oil spill behavior model. Pp. 525-539 in *Proceedings of the 23rd Arctic and Marine Oil Spill Program (AMOP) Technical Seminar, Vancouver, BC*. Environmental Protection Service, Environment Canada, Ottawa, Ontario.

Lindstrom, J.E., and J.F. Braddock. 2002. Biodegradation of petroleum hydrocarbons at low temperatures in the presence of dispersant Corexit 9500. *Marine Pollution Bulletin* 44:739-747.

Mackay, D., S. Paterson, and K. Trudel. 1980. *A Mathematical Model of Oil Spill Behavior*. Department of Chemical and Applied Chemistry, University of Toronto, Canada.

Marty, G.D., J.W. Short, D.M. Dambach, N.H. Willits, R.A. Heintz, S.D. Rice, J.J. Stegeman, and D.E. Horton. 1997. Ascites, premature emergence, increased gonadal cell apoptosis, and cytochrome P4501A induction in pink salmon larvae continuously exposed to oil-contaminated gravel during development. *Canadian Journal of Zoology* 75:989-1007.

Mooers, C.N.K., and J. Wang. 1998. On the implementation of a 3-D circulation model for Prince William Sound, Alaska. *Continental Shelf Research* 18:253–277.

National Research Council (NRC). 2002. *Oil in the Sea: Inputs, Fates, and Effects*. National Academy Press, Washington, DC.

National Research Council (NRC). 1978. *OCS Oil & Gas: An Assessment of the Department of the Interior Environmental Studies Program*. National Academy Press, Washington, DC.

OSRI (Oil Spill Recovery Institute). 2002a. OSRI Annual Work Plan FY02. Oil Spill Recovery Institute, Cordova, AK. Also available online at http://www.osri-pwssc.org.

OSRI (Oil Spill Recovery Institute). 2002b. OSRI Grant Policy Manual. Oil Spill Recovery Institute, Cordova, AK. Also available online at http://www.osri-pwssc.org.

OSRI (Oil Spill Recovery Institute). 2002c. *Science and Environmental Education Program in South-Central Alaska - Forging a new Alliance*. Oil Spill Recovery Institute, Cordova, AK.

OSRI (Oil Spill Recovery Institute). 2002d. *Technology Coordinator's Report*. Oil Spill Recovery Institute, Cordova, AK.

OSRI (Oil Spill Recovery Institute). 2001. *OSRI Annual Work Plan FY01*. Oil Spill Recovery Institute, Cordova, AK. Also available online at http://www.osri-pwssc.org.

OSRI (Oil Spill Recovery Institute). 2000. *OSRI Annual Work Plan FY00*. OSRI Business Plan. Oil Spill Recovery Institute, Cordova, AK. Also available online at http://www.osri-pwssc.org.

OSRI (Oil Spill Recovery Institute). 1999. *OSRI Business Plan*. Oil Spill Recovery Institute, Cordova, AK. Also available online at http://www.osri-pwssc.org.

OSRI (Oil Spill Recovery Institute). 1998. *OSRI Grant Policy Manual*. Oil Spill Recovery Institute, Cordova, AK.

OSRI (Oil Spill Recovery Institute). 1995. *Oil Pollution and Technology Plan for the Arctic and Subarctic*. Oil Spill Recovery Institute, Cordova, AK.

OSRI (Oil Spill Recovery Institute). Bylaws. Oil Spill Recovery Institute, Cordova, AK. Also available online at http://www.osri-pwssc.org.

Parker, W. 2002. OSRI Advisory Board. Personal communication, 1st meeting briefing, "Background on OSRI Formation and Purpose," February 7, 2002, Anchorage, Alaska.

Payne, J.R., B.E. Kirstein, J.R. Clayton, Jr., C. Clary, R. Redding, G.D. McNabb, Jr., and G. Farmer. 1987. *Integration of Suspended Particulate Matter and Oil Transportation Study, Final Report*. Minerals Management Service, Environmental Studies Branch, Contract No. 14-12-0001-30146, Anchorage, Alaska.

Payne, J.R., Jr., J.R. Clayton, Jr., G.D. McNabb, B.E. Kirstein, C.L. Clary, R.T. Redding, J.S. Evans, E. Reimnitz, and E.W. Kempema. 1989. Oil-ice-sediment interactions during freezeup and breakup. Pp. 1-382 in *Outer Continental Shelf Environmental Assessment Program, Final Reports of Principal Investigators*. U.S. Department of Commerce, National Oceanic and Atmospheric Administration, OCSEAP Final Rep. Final Rep. 64, Washington, DC.

Public Law 101-380. Aug. 17, 1990, Title V. Prince William Sound Provisions. Available online at <http://www.pwssc-osri.org/docs/opa90.html>.

Reed, M. 2002. Technical Description and Verification Tests of OSCAR 2000: A multi-component 3-dimensional oil spill contingency and response model. SINTEF Applied Chemistry, Environmental Engineering, April 2002. Trondheim, Norway.

Reed, M., O.M. Aamo, and P. Daling. 1995. OSCAR, a model system for quantitative analysis of oil spill response strategies. Pp. 815-835 in *Proceedings of the 18th Arctic Marine Oilspill Program (AMOP) Technical Seminar*. Environment Canada, Edmunton, Alberta.

Reed, M., P.S. Daling, O.G. Brakstd, I. Singsaas, L.-G. Faksness, B. Hetland, and N. Efrol. 2000. OSCAR 2000: A multi-component 3-dimensional oil spill contingency and response model. Pp. 663-680 in *Proceedings of the 23rd Arctic Marine Oilspill Program (AMOP) Technical Seminar*. Environment Canada, Ottawa, Ontario.

Reed, M., T. Kana, and E. Gundlach. 1988. Development, testing and verification of an oil spill surf-zone mass-transport model. Final Report to Mineral Management Service, Alaska OCS Region, Contract No. 14-12-0001-30130; by Applied Science Associates, Inc. (ASA), Coastal Science & Engineering, Inc. (CSE), and E-Tech, Inc., June 1988.

Reed, M., E. Gundlach, and T. Kana. 1989. A coastal zone oil spill model: development and sensitivity studies. *Oil & Chemical Pollution* 5:411-449.

Robertson, T.L., and E. DeCola. 2001. Final Proceedings of the Prince William Sound Meteorological Workshop, held December 12-14, 2000 in Anchorage, Alaska. Oil Spill Recovery Institute, Cordova, Alaska.

Roy, N.K., J. Stabile, J. E. Seeb, C. Habicht, and I. Wirgin. 1998. High frequency of K-*ras* mutations in pink salmon embryos experimentally exposed to *Exxon Valdez* oil. *Environmental Toxicology and Chemistry* 18(7):1521-1528.

Thomas, G.L., and R.E. Thorne. 2001. Night-time predation by Steller sea lions. *Nature* 411:1013.

Thomas, G.L., and W.D. Cox. 2000. A nowcast-forecast information system for Prince William Sound. Pp. 247-255 in *Proceedings of the 23rd Arctic and Marine Oil Spill Program (AMOP) Technical Seminar June 14-16, 2000, Volume I*. Environment Canada, Vancouver, British Columbia.

U.S. Congress, House. 1996. U.S. Coast Guard Authorization Act. H.R. 854. 104th Cong., 2d. session, Appropriations Authorization, pp. 4-6.

Wang, J. 2001. A nowcast/forecast system for coastal ocean circulation using simple nudging data assimilation. *Journal of Atmospheric and Oceanic Technology* 18(6):1037-1047.

White, P.A., S. Robitaille, and J.B. Rasmusssen. 1999. Heritable reproductive effects of benzo [a]pyrene on the fathead minnow (*Pimephales promelas*). *Environmental Toxicology and Chemistry* 18(8):1843-1847.

Wolfe, D.A., M.J. Hameedi, J.A. Galt, G. Watabayashi, J. Short, C. O'Clair, S. Rice, J. Michel, J.R. Payne, J. Braddock, S. Hanna, and D. Sale. 1994. The fate of the oil spilled from the Exxon Valdez. *Environmental Science and Technology* 28:561A-568A.

Appendixes

A

Committee Biosketches

Dr. Mahlon C. Kennicutt II, *Chair,* is director of the Geochemical and Environmental Research Group (GERG), adjunct professor of oceanography, member of the toxicology faculty, and team leader of the Sustainable Coastal Margins Program (SCMP) at Texas A&M University. Dr. Kennicutt earned his Ph.D. in oceanography in 1980 from Texas A&M University. His research interests include environmental monitoring; fate and effects of contaminants; environmental impacts of offshore energy exploration and exploitation; coordination of the social and physical sciences to address environmental issues; and all aspects of sustainable development of coastal margins. He is currently an ex officio member of the Polar Research Board and is the U.S. delegate to the Scientific Committee on Antarctic Research. He previously served on the PRB's Committee to Review NASA's Polar Geophysical Data Sets and is serving on the current committee Cumulative Environmental Effects of Oil and Gas Activities on Alaska's North Slope. Dr. Kennicutt is a member of various professional organizations including the American Geophysical Union, the American Association for the Advancement of Science, and the American Society of Limnology and Oceanography.

Dr. Brenda Ballachey is a research physiologist with the U.S. Geological Survey in Anchorage, Alaska. Dr. Ballachey earned her Ph.D. in animal breeding and genetics from Oregon State University in 1985. Her areas of expertise are marine mammals (population status and indexes of condition); sea otters (biochemical, physiological, population and ecological effects of oil exposure); environmental toxicology; and biomarkers of contaminant exposure. Dr. Ballachey brings to the committee her long-term experience as a project leader for population status and sea otter oil spill studies and as a principal investigator on the Nearshore Vertebrate Predator project, an ecosystem approach to examining the recovery of the coastal marine environment in Prince William Sound, Alaska, after the 1989 *Exxon Valdez* oil spill.

Dr. Joan Braddock is a professor of microbiology at the University of Alaska, Fairbanks. Dr. Braddock earned her Ph.D. in oceanography from the University of Alaska, Fairbanks. Her research includes extensive work on biodegradation of oil in intertidal and subtidal marine sediments following the *Exxon Valdez* oil spill; evaluation of bioremediation as a treatment technology following the *Exxon Valdez* oil spill; bioremediation effectiveness in Arctic and subarctic terrestrial ecosystems; natural attentuation of contaminants in cold climates; the effect of sediments on bioavailability of petroleum hydrocarbons for microbial degradation; and microbial degradation of chlorinated and unchlorinated hydrocarbons in groundwater.

Dr. Akhil Datta-Gupta is an associate professor and holder of Robert L. Adams Professorship in Petroleum Engineering at Texas A&M U. in College Station. He has a Ph.D. in petroleum engineering from the University of Texas at Austin. Previously, he worked with BP Exploration at Alaska and also at the BP Research and the Lawrence Berkeley National Laboratory. He has extensive experience in subsurface characterization and fluid flow modeling, both for oil recovery and environmental remediation. His research interests include high-resolution numerical schemes for reservoir simulation; geostatistics and stochastic reservoir characterization; modeling and scale-up of enhanced oil recovery; environmental remediation and contaminant transport. He is a distinguished member of the Society of Petroleum Engineers and currently serves on the NRC's Polar Research Board.

Dr. Deborah French McCay is a principal at Applied Science Associates, Inc. Dr. French earned her Ph.D. in biological oceanography in 1984 from the University of Rhode Island. Her research interests include quantitative assessments and modeling of aquatic ecosystems to assess environmental impact and damage, especially in regards to response to pollutants, including oil. She is an expert in modeling oil fates and effects, toxicity, exposure to bioaccumulation of pollutants by biota, along with the effects of this contamination. Her models have been used for impact, risk, and natural resource damage assessments, as well as for studies of biological systems. Her expert testimony has been provided in hearings regarding environmental risk and impact assessments.

Dr. Jerry Neff is a senior research leader at the Battelle Memorial Institute. Dr. Neff earned his Ph.D. in zoology in 1967 from Duke University. He has performed extensive research for the oil industry and the U.S. federal government on the aquatic environmental fate and effects of heavy metals and petroleum hydrocarbons from offshore drilling and production operations, clean ballast water discharges from tankers, and from major oil spills, including the *Esso Bayway* crude oil spill in the Neches River, the Arthur Kill, New Jersey diesel spill, *Amoco Cadiz* crude oil spill in France, the *Exxon Valdez* crude oil spill in Alaska, the Newton Lake, Illinois, pipeline break and oil spill, the *Haven* oil spill off Genoa, Italy, the Trecate oil spill in rice fields north of Milan, Italy, and the *Seki* oil spill in the United Arab Emirates. He has been a member of three review panels of the National Research Council, the first dealing with Fate and Effects of Drilling Mud and Cuttings in the Marine Environment and the second dealing with marine oil spills. The third was an assessment of marine environmental monitoring in the Southern California Bight.

Dr. James Payne is the president of Payne Environmental Consultants, Inc. Dr. Payne earned his Ph.D. in chemistry in 1974 from the University of Wisconsin-Madison, and he was a Woods Hole Oceanographic Institution postdoctoral scholar from 1974 to 1975. Payne Environmental Consultants, Inc., specializes in oil and chemical pollution studies for government and industry. Over the 28 years of his professional career, Dr. Payne has been involved in numerous projects dealing with marine- and water-pollution issues, including laboratory and flow-through wave-tank studies of oil weathering behavior in Arctic and subarctic waters. He has also supported NOAA natural resource damage assessment efforts after the *Exxon Valdez*, *American Trader*, *Kuroshima*, and *New Carissa* oil spills. Dr. Payne contributed background chapters for the 1985 NRC publication *Oil in the Sea—Inputs, Fates, and Effects* and he was a member of the NRC Ocean Science Board committees dealing with the Effectiveness of Oil Spill Dispersants (1985-1988) and Spills of Emulsified Fuels (2001).

Dr. James P. Ray is a manager of environmental ecology and response at Shell Global Solutions, Inc. Dr. Ray earned his Ph.D. in biological oceanography in 1974 from Texas A&M University. His research interests include petroleum industry marine ecological research on fate and effects contaminants in the marine environment and related technology, and he is a well-known inputs expert. Dr. Ray was previously a member of the NRC Committee to Review Fate and Effects of Drilling Fluids in the

Marine Environment; the Committee on Arctic Marine Sciences; and Managing Troubled Waters: The Role of Marine Environmental Monitoring. He also served on the NRC steering group for "Planning a Systems Assessment of Marine Environmental Monitoring."

Dr. Bill Sackinger is the president and CEO of OBELISK Hydrocarbons, Ltd., in Fairbanks, Alaska. Dr. Sackinger earned his Ph.D. in electrical engineering from Cornell University in 1969. His research interests include sea ice dynamics; naturally induced stresses in sea ice sheets near grounded obstacles; theoretical analysis of artificially formed sea spray ice-freezing kinetics; and research on sea ice failure stresses. Dr. Sackinger has provided consulting and advisory experience to several state, federal, and international agencies, including the State of Alaska, U.S. Department of Energy, and the Japan National Oil Corporation. He has also provided consultation for the NRC's Polar Research Board on various occasions since 1974.

B

OSRI Bylaws

Bylaws of the
Prince William Sound
Oil Spill Recovery Institute
(as revised November 2001)

Table of Contents

Article I. Definitions

<u>Board</u> means the Advisory Board of the Prince William Sound Oil Spill Recovery Institute.

<u>Chairperson</u> is the representative of the Secretary of Commerce who serves as Chair of the Advisory Board.

<u>Director</u> is the Oil Spill Recovery Institute's Director, appointed by the Advisory Board. The Scientific and Technical Committee and the Prince William Sound Science and Technology Institute (dba Prince William Sound Science Center) may each submit independent recommendations for the Advisory Board's consideration for appointment as Director. The Director may hire such staff and incur such expenses on behalf of the Institute as are authorized by the Advisory Board.

<u>Institute</u> means the Prince William Sound Oil Spill Recovery Institute

<u>Center</u> means the Prince William Sound Science Center, a non-profit organization incorporated as the Prince William Sound Science and Tech-

nology Institute. This organization is authorized by the Secretary of Commerce to administer the Institute.

Article II. Identification and Offices

Section 1.   Name. The name of the Institute is the Prince William Sound Oil Spill Recovery Institute.

Section 2.   Offices. The offices of the Institute shall be located in Cordova, Alaska.

Article III. Purposes

The Prince William Sound Oil Spill Recovery Institute was authorized by Section 5001 of the Oil Pollution Act of 1990, approved by the United States Congress in August 1990 and amended by Public Law 101-380 in the Coast Guard Reauthorization Act of 1996. The authorizing legislation as amended states that "The Secretary of Commerce shall provide for the establishment of a Prince William Sound Oil Spill Recovery Institute (hereinafter in this section referred to as the "Institute") through the Prince William Sound Science and Technology Institute located in Cordova, Alaska."

Section 1.   Functions. The Institute will conduct research and carry out educational and demonstration projects designed to:

> (A)   identify and develop the best available techniques, equipment, and materials for dealing with oil spills in the Arctic and subarctic marine environment; and

> (B)   complement Federal and State damage assessment efforts and determine, document, assess and understand the long-range effects of Arctic or subarctic oil spills on the natural resources of areas affected by these spills, and the environment, the economy and the lifestyle and well-being of the people who are dependent on them.

Section 2.   Prohibited Activities. No Congressionally appropriated funds may be used for the acquisition of land or buildings. No Congressionally appropriated funds may be used to initiate litigation.

Section 3.   Termination. The Institute shall terminate in August of
the year 2006, as stipulated by the 1996 amended Section 5001(i) of the Oil
Pollution Act of 1990, or at any other date as determined by Congress.

Article IV. Advisory Board

Section 1.   Policies. Policies of the Institute shall be determined by
the Advisory Board.

Section 2.   Establishment. The Secretary of Commerce shall establish
the Institute through the Center, as stipulated in Section 5001(a) of the Oil
Pollution Act of 1990.

Section 3.   Fiscal Year. The fiscal year of the Institute shall coincide
with the Federal Government's, beginning October 1 and ending September 30.

Section 4.   Composition of the Board. Appointments to the Board are
made as follows:

> (A)   One representative appointed by each of the
> Commissioners of Fish and Game, Environmental
> Conservation, and Natural Resources, of the State
> of Alaska, all of whom shall be State employees.
>
> (B)   One representative appointed by each of the
> Secretaries of Commerce, the Interior, and Trans-
> portation, who shall be Federal employees.
>
> (C)   Two representatives from the fishing indus-
> try appointed by the Governor of the State of Alaska
> from among residents of communities in Alaska that
> were affected by the EXXON VALDEZ oil spill, who
> shall serve terms of two years each. Interested orga-
> nizations from within the fishing industry may sub-
> mit the names of qualified individuals for consider-
> ation by the Governor.
>
> (D)   Two Alaska Natives who represent Native
> entities affected by the EXXON VALDEZ oil spill, at
> least one of whom represents an entity located in
> Prince William Sound, appointed by the Governor

of Alaska from a list of four qualified individuals submitted by the Alaska Federation of Natives, who shall serve terms of two years each. Nominations from all Alaska Native organizations will be encouraged.

(E)   Two representatives from the oil and gas industry to be appointed by the Governor of the State of Alaska who shall serve terms of two years each. Interested organizations from within the oil and gas industry may submit the names of qualified individuals for consideration by the Governor.

(F)   Two at-large representatives from among residents of communities in Alaska that were affected by the EXXON VALDEZ oil spill who are knowledgeable about the marine environment and wildlife within Prince William Sound, and who shall serve terms of two years each, appointed by the remaining members of the Advisory Board. Interested parties may submit the names of qualified individuals for consideration by the Advisory Board;

(G)   One non-voting representative appointed by the Institute of Marine Science.

(H)   One non-voting representative appointed by the Prince William Sound Science & Technology Institute (dba as the Prince William Sound Science Center).

Section 5.   Terms of Office. The terms of office for the at-large representatives and those appointed by the Governor shall be two years from their dates of appointment. All other representatives serve until replaced by the Governor, agency or organization they represent.

Section 6.   Alternates. Appointed members may send an alternate to meetings they are unable to attend, but the alternate will not have full voting privileges. Alternates designated in writing (either via letter or e-mail) by the appointed member for a specific meeting will be seated and may fully participate in discussions but votes may only be cast by the officially appointed representative. This applies to all voting members.

Section 7.   Resignations and Vacancies. Resignations of any Board members shall be made in writing to the Board Chairperson and their agency or appointing organization. OSRI shall seek nominations, as appropriate, to fill vacancies; these appointments will be made to complete the unexpired term of the vacated Board seat. Vacancies of Board members representing state and federal government agencies, the Institute of Marine Science and the Prince William Sound Science Center will be appointed by the agency or organization.

Section 8.   Organization of the Institute. The Advisory Board will appoint the Director after considering recommendations received from the Center and the Scientific and Technical Committee. The Director may hire staff and incur expenses as authorized by the Board.

Section 9.   Fiduciary Responsibilities. As the designated administrator for the Institute, the Center and its Board of Directors are responsible to maintain accurate accounting of the Institute's funds and to disperse monies only as approved by the Institute's Advisory Board. Institute funds shall be maintained in separate accounts from other Center funds. Annual audits of the Center's financial records shall be conducted and copies made available to the Advisory Board.

Section 10.   Conflicts of Interest. A conflict of interest may arise where there exists the opportunity for direct or indirect material personal gain by a Board member. Any duality on the part of any Board member or a member of the Board member's family shall be immediately disclosed to the Advisory Board, and made a matter of record in the minutes. Members of the Advisory Board shall complete a Conflicts of Interest Disclosure form, in substantially the form as shown in Appendix A, upon becoming a Board member and whenever a potential conflict of interest arises. Any application for a contract, grant or award shall include a Conflicts of Interest Disclosure form from each of the Principal Investigator (PI) and Authorized Agent (AA) to assist the OSRI Director in avoiding conflicts throughout the review process.

Any Board member having a duality of interest shall not vote or participate in a peer review of the matter. Such Board member shall be counted in determining the quorum for a meeting, however. The minutes of the meeting shall reflect that a disclosure was made, and the abstention from participating in voting. The Board member shall refrain from any discussion unless questioned by the Board.

Any grant, contract or award in which a Board member or a member of the Board member's family has an interest shall be valid notwithstanding such interest, if the material facts of such interest are disclosed or known to the Board in advance, and the Board shall nevertheless, authorize and ratify the grant, contract or award, PROVIDED THAT, no Board member shall supervise a family member or administer a grant, contract or award to a family member.

In addition to the above, Board members are obliged to comply with any applicable federal, state, municipal or organization requirements regarding conflict of interest and disclosure, and political activities.

Article V. Officers of the Advisory Board

Section 1.   Officers. The officers of the Board are a Chairperson, a Vice Chairperson, a Secretary and a Treasurer.

Section 2.   Chairperson. The representative of the Secretary of Commerce shall serve as Chairperson of the Board.
(

> A)   The Chairperson shall preside at meetings of the Board.

> (B)   Except as authorized by resolutions of the Board, the Chairperson may sign all contracts, resolutions, and other instruments made by the Board.

Section 3.   Vice Chairperson. The Vice Chairperson shall perform the duties of the Chairperson in the absence or incapacity of the Chairperson. The Board shall elect the Vice Chairperson for a two-year term.

Section 4.   Secretary. The Board shall elect the Secretary for a two-year term. The duties of the Secretary shall include:

> (A)   The Secretary is responsible for the minutes of all Board meetings. Staff will assist the secretary as required.

> (B)   The Secretary will draft correspondence, resolutions or other documents as authorized by the Board.

> (C)   The Secretary will make sure credentials of Board members are in order.

(D)   Performing other duties as assigned from time to time by the Board.

Section 5.   Treasurer. The Board shall elect the Treasurer for a two-year term. The duties of the Treasurer shall include:

(A)   Oversight and review of the audit and recordkeeping functions of the Institute;

(B)   Review of the financial and annual plans of the Institute;

(C)   Oversight and review of the staff's disbursements of funds for Institute grants, contracts, expenses and other obligations;

(D)   Presentation of financial reports to the Board;

(E)   Performing other duties as assigned from time to time by the Board.

Section 6.   Elections. Officers shall be elected every two years or as necessary to fill vacancies. The two additional members of the Executive Committee shall be elected annually as necessary, with no term limits.

Section 7.   Vacancies of officers. If the Vice Chairperson, Secretary or Treasurer's office becomes vacant, the Board shall elect a successor from among remaining Board members.

Section 8.   Process of electing officers. Nominations shall be received from the floor for the three elective officers by any member of the Board. Any number of nominations may be made. After nominations have been closed, a secret ballot will be held. Officers shall be elected by a simple majority; if there is no simple majority there shall be a run-off election between the two candidates receiving the most votes. Only voting members shall be allowed to vote.

Article VI. Meetings

Section 1.   Annual Meeting. The annual meeting of the Board shall be held in the fall of each year at the place and time and on the dates fixed by the Chairperson by telegraph, telefax or written notice to the members of the Board at least twenty (20) days prior to the date of such annual meeting. The annual meeting is also a meeting as set forth in Section 2 of this Article.

Section 2.   <u>Meetings.</u> Meetings of the Advisory Board shall be held in the spring and fall of each year on the date and at the time and place designated at the last regular meeting. In the absence of such designation, then the meeting shall be held on the date and at the time and place in any such month as fixed by the Chairperson.

> (A)   Meetings will be held to review and approve policies for the conduct and support, through contracts and grants awarded on a nationally competitive basis of research, projects and studies to be supported by the Institute in accordance with the purposes of the amended Section 5001 of the Oil Pollution Act of 1990.

> (B)   Members of the Board shall have at least twenty (20) days prior notice of meetings; designation of date, time and place of a meeting at the previous meeting constitutes sufficient prior notice as proscribed in Section 1 of this article.

> (C)   If a waiver of Notice and Consent of the absent members is provided, then any and all business may be transacted notwithstanding the nonprovision of prior notice of a meeting to Board members.

> (D)   Any Board member who misses two consecutive meetings will be queried by the Secretary as to their ability for full participation; depending on the response and their interest, appropriate action will be taken.

Section 3.   <u>Special Meetings.</u> The Chairperson of the Board may, when he/she deems it expedient, and shall upon the written request of at least eight Board members, call a special meeting of the Board for purpose of transacting any business designated in the call.

> (A)   The call for a special meeting shall be wired, faxed or mailed to the business or home address of

members of the Board at least five (5) working days prior to the date of such special meeting.

(B)   At such special meetings, no business shall be considered other than as designated in the call.

i) **Exception.** If all Board members are present at a special meeting, or those not present have signed a waiver of Notice and Consent to Meeting, a quorum otherwise being present, any and all business may be transacted at such special meeting.

(C)   Public notice for a special meeting shall be given pursuant to Section 5 of this Article.

Section 4.   Teleconference meetings. Any meeting of the Advisory Board (regular, special or the annual meeting) may be held by teleconference.

Section 5.   Public Notice. All meetings of the Board shall be preceded by reasonable notice to the public of the time, place and subject matter of the proceeding.

Section 6.   Quorum. A quorum shall be constituted by a majority of the voting members of the Board.

(A)   When a quorum is present, action may be taken by the Board only upon the affirmative vote of a majority of the members present.

Section 7.   Rules of Procedure. The rules of procedure applicable at all regular, special and committee meetings of the Board are Robert's Rules of Order of the most recent edition, except as the Board provides by resolution for other procedures.

Section 8.   Voting and Resolutions. Each member of the Board in attendance at a meeting of the Board shall have the right to cast one vote on any question brought before such body during a meeting with the exception of the two non-voting Board members representing the Institute of Marine Science and the Prince William Sound Science Center.

(A)   Only those members present at a meeting,

either physically at the meeting or participating by teleconference, may vote.

(B) Alternate members have the same privileges as non-voting members.

(C) At the request of a Board Member a roll call vote will be taken.

(D) For purposes of this section, "present" shall mean the Board member is either physically in the room where the meeting is being held or, the member is participating in the meeting via teleconference as provided for in Section 4.

(E) Non-voting members shall have the right to introduce and second motions to the floor including nominations for Board Officers.

Section 9. Executive Sessions. The Board may retire into executive session at any time in compliance with applicable federal, state, municipal or organizational regulations regarding executive session.

Article VII. Committees

Section 1. Executive Committee. This committee shall be composed of the Chair, Vice-Chair, Secretary, Treasurer and two additional members of the Advisory Board.

Primary purposes of the Executive Committee are to:

1. Assist and supervise implementation of the OSRI work plan between meetings of the Advisory Board, and
2. Assist in developing partnerships with private industries and other public entities.

Subject to limitations imposed by the Board of Directors, the Executive Committee shall exercise the authority of the Advisory Board, **except that** the Executive Committee shall **not** have authority to:

1. Amend the bylaws.
2. Adopt a merger or consolidations plan with another organization.

3.   Authorize the sale, lease, exchange or mortgage of all or substantially all of the property or assets for the Institute.

4.   Modify any of the individual program budgets—i.e., Technology, Ecology, Education or Administration—by more than twenty percent.

Meetings of the Executive Committee are scheduled by the Chair or Vice-Chair. Any Advisory Board member or the Director may request an Executive Committee meeting be scheduled with notice as detailed in Article VI.

The Executive Committee shall meet on at least an annual basis to review performance of the Director.

Section 2.   Scientific and Technical Committee. The Board shall establish a scientific and technical committee, composed of specialists in matters relating to oil spill containment and cleanup technology, Arctic and subarctic marine ecology, and the living resources and socio-economics of Prince William Sound and its adjacent waters. These specialists will come from the University of Alaska, the Institute of Marine Science, the Prince William Sound Science Center and elsewhere in the academic community. The "academic community" includes all individuals identified by the Board, regardless of affiliation, for the expertise they may lend in basic and applied research and development.

>(A) Function of the committee: The Scientific and Technical Committee shall provide such advice to the Board through the Director of the Institute including recommendations regarding the conduct and support of research, projects, and studies in accordance with the purposes of this section. The Board shall not request and the Committee shall not provide any advice on damage assessment which is not directly related to Arctic or subarctic oil spills or the effects thereof.

>(B) Committee Chairperson: This Committee Chairperson shall be appointed by the Board Chairperson and shall serve a two-year term, or otherwise serve at the pleasure of the Board Chairperson.

(C) Review of proposals: The Scientific and Technical Committee shall review proposals in accordance with the process identified in the OSRI Grant Policy Manual.

Section 3.   Other committees. The Board shall appoint other committees as necessary to complete the objectives of the Institute.

Article VIII. Amendments

These Bylaws may be altered or amended at any duly organized meeting of the Advisory Board by a two-thirds majority vote of the Board members then serving in office.

Appendix A - OSRI Bylaws
Oil Spill Recovery Institute
<u>Conflicts of Interest Disclosure Form</u>

This form is being completed in accordance with Article IV, Section 9 of the OSRI Bylaws and Section 8.2.2 of the OSRI Grant Policy Manual.

1.   Please list all business and professional activities in which you or any immediate family hold an interest as owner (excluding stock ownership in a publicly traded corporation), officer, board member, partner, employee or other beneficiary position.

2.   Please identify any family or business relationship with any current Board member or employees of OSRI and identify the names of the persons, the positions they hold, and the nature of the relationship.

3.   Provide a brief description of all legal proceedings pending or threatened to which you or another person with whom you are affiliated or a family member are a party with interests adverse to OSRI.

4.   Do you or any member of your family have an arrangement or understanding for employment by any recipient of a current grant or award by OSRI or are there any future financial transactions anticipated between OSRI and such persons or entities? If so, please describe the terms and parties to each such arrangement or understanding.

I certify that the above information is true and correct to the best of my knowledge, that I am not in a position of possible conflict of interest except as stated above or as follows:

and that I will promptly disclose any potential conflicts to the Advisory Board as they arise.

Date:

Signature:

Printed Name:

C

OSRI Advisory Board Meeting Minutes

Wednesday, August 12, 1998

The meeting was held in the library meeting room, Cordova, Alaska.

Board members present:
Barbara Moore, Department of Commerce
Doug Mutter, Department of Interior
Cmdr. Ross Tuxhorn, Department of Transportation
R.J. Kopchak, Fishing industry representative
Grant Vidrine, Oil and gas industry representative
Larry Dietrick, Alaska Department of Environmental Conservation
Glenn Ujioka, Native representative
Gail Evanoff, Native representative
Marilyn Leland, At-large representative
Mead Treadwell, non voting representative, Prince William Sound Science Center
John Goering, non voting representative, Institute of Marine Science, Univ. of Alaska, Fairbanks

Not present: Governor Steve Cowper, Kevin Meyers, Carol Fries, Claudia Slater, and Virginia Adams.

Staff and visitors: Gary Thomas, Penny Oswalt (teleconference), Nancy Bird, Maxwell Blair, Liz Senear, Bruce Wright (EVOS), Vince Patrick, Jay Kirsch, Switgard Duesterloh.

Advisory Board Chair Barbara Moore called the meeting to order at 8:05 a.m. and asked for introductions by all present.

Approval of agenda: Thomas would like discussion of indirect costs added. Treadwell would like to report on Arctic Research Association. **Motion** by Leland, seconded by Vidrine, to accept agenda with those additions. **Motion passed.**

Public comments: None.

Approval of minutes: **Motion** by Kopchak, seconded by Leland to accept minutes. **Amended** by Leland to include full language of bylaw amendments passed at the 3/16/98 meeting. Friendly amendment accepted. **Motion passed.**

Chair's report: Moore reported on the legal issues outstanding, which both the PWSSC legal staff, and the NOAA legal staff have looked at and written opinions about. These issues are 1) whether the Technology Coordinator position is an administrative or program expense, 2) whether projects under a certain dollar amount need to be advertised nationally, and 3) what are the rules governing the use of payback money. Moore reported that the NOAA general counsel says the payback money is not considered to be U.S. Gov't. funds and is not subject to treasury rules and regulations. It is up to OSRI, so OSRI should write rules governing the use of payback money. Moore suggested that during the course of a project the payback money go back to the project, and after the project is completed, it goes back to OSRI to be used for that program.

Discussion   Treadwell requested a briefing on the intellectual property policy found in the grants manual, and Vidrine requested a summary of OMB circular A-110 as referenced in section 10.4.1 of the grant policy manual. The draft OSRI policy states that the grantee retains 10% of the money and the rest goes back to OSRI. Concerns were expressed that the policy was not realistic enough, that there needs to be more flexibility in grant negotiation, and that the policy needs to be delineated more specifically.

MOTION	**Motion** by Moore, seconded by Kopchak that the Board accept the present policy on payback funds, with the staff directed to research this issue in more depth and present a report to the Board. **Motion passed.**
Action Item	Moore recommends that a Memorandum of Understanding (MOU) between the Dep't. of Commerce and PWSSC regarding administration of OSRI be drafted and signed. She will send OSRI staff a draft MOU that the Dep't of Commerce uses with other organizations. Treadwell suggests considering other possible ties between NOAA and OSRI.
MOTION	**Motion** by Moore, seconded by Kopchak that the staff draft a MOU for consideration at the next Board meeting, taking into account other possible links between OSRI and NOAA programs. **Motion passed.**
Technology Program Coordinator	The NOAA general counsel says that the legislation does not prohibit this being treated as a program expense but the Board would have to make an explicit decision to do so.
MOTION	**Motion** by Kopchak, seconded by Leland, that program funds be used to hire the technology coordinator, and that the position be an OSRI staff position. **Motion passed.**
Grant advertising	A discussion ensued about whether all projects have to be advertised nationally, regardless of the dollar amount of the award. Mutter says that the legal opinion by Brookings seems to say that for smaller grants OSRI doesn't have to advertise nationally, but Moore says that this conflicts with the opinion of NOAA's general counsel. Treadwell suggested issuing a Broad Area Announcement (BAA) once a year which is essentially a published policy statement which serves as a request for proposals (RFP), and covers everything. Moore pointed out that the federal system does something similar. Resolving this issue may require further legal counsel. The discussion was deferred until the afternoon, when Bird will have procured the U.S. code.

Arctic Research Commission

Treadwell reported that this presidentially appointed commission is looking at U.S. policies and research in the Arctic. At the last meeting OSRI was urged to do everything possible to solve in situ burning impasses which has prevented Clean Seas from proceeding with broken ice issues on the North Slope. The U.S. will take on council leadership in September which is an opportunity for Alaskans doing Arctic research. OSRI is being looked to as a leader. There was strong interest in SEA type research being expanded to other areas. Thomas said Gov. Knowles is putting together a science panel to discuss Bristol Bay and would like help from OSRI regarding a SEA type program.

Executive Director's Report

Action List Update: Thomas briefed the Board on the status of the Action List from the March 16 meeting, as summarized in the meeting packet. *Sensitive resources mapping*: the report from Whitney (NOAA) is a summary of what is going on in GIS mapping in the U.S. According to Mutter it includes both public agency and private company work. Kopchak requested a copy, and Moore suggested making copies available to Board members. There is an effort to get all EVOS information on the GIS network. *Business Plan Review*: Thomas discussed the comments Dietrick had submitted regarding the business plan. Dietrick said that OSRI needs to play a coordinating role in finding out what is going on with current research in Alaska. Dietrick has already taken some action in this direction. He would also like OSRI to look into ways to integrate the technology that is/has been developed by OSRI with ongoing science. Thomas would like Board members to comment on material that has been compiled on these topics. The field of oil pollution is primarily finding engineering/technology solutions to pollution problems, but one can't know how to appropriately use the technology if one doesn't understand the ecosystem. *Science planning workshop*: Thomas is negotiating with 4 groups regarding proposals. He would like to see more horizontal/vertical integration including users, managers, and researchers than most of the proposals show.

Grants   Thomas gave a short summary of the contracts listed
Programs   in Table I. Treadwell mentioned that he thought OSRI
                  should electronically publish along with standard pub-
                  lishing. He says there are firms in Alaska who will put
                  a publication into an electronically reproducible for-
                  mat for $150. Mutter suggested getting a standard dis-
                  tribution list together for any OSRI publications.

Action Item   It was decided that the Proceedings of the Symposium
                  on Practical Ice Observations should be more widely
                  distributed. RCAC printed this, and there are still
                  some funds left. Thomas was directed to interface with
                  Mutter to determine how many more copies were
                  needed for a re-print. Some should be mailed to the
                  list produced and some kept on hand for requests.
                  Vidrine would like a copy.

                  Brochure Stan Stephens' boats, the state ferries, and
                  the communities were all suggested for dispersal of
                  brochures. Blair has given some to cruise ships, and
                  Bird plans to send some to the Congressional offices in
                  Washington, D.C. and Anchorage. Treadwell sug-
                  gested the Inter-Agency Committee's mailing list.

                  Website: Bird asked if it is appropriate to place unap-
                  proved minutes on the website. No objections. It was
                  recommended to simplify the website URL to some-
                  thing like osri.com. Treadwell suggested OSRI explore
                  becoming a member of ARLISS.

                  Thomas presented a short summary of the grant pro-
                  posals as listed in the FY99 Draft business plan.

Technology   The Nowcast/Forecast circulation model has gone
                  through committee review and will be discussed when
                  the science committee presents its report. OSRI has an
                  obligation to provide the Arctic Council digital infor-
                  mation for the circumpolar map for resources at risk
                  and oil distribution. There is $30,000 allotted for this,
                  but a BAA has not yet been issued. $20,000 for a work-
                  shop on oil recovery in broken ice has been set aside
                  but a BAA has not yet been issued. OSRI is waiting for
                  a funding partner. There is $45,000 in uncommitted

funds in the technology program. Dietrick commented that regarding in situ burning in broken ice, current proposals were turned down by ADEC because it was work that has already been done. Alaska Clean Seas is redesigning their experiments.

Ecology

One half of the Nowcast/Forecast funds will come from the ecology program. There is a proposal from Kline of PWSSC to do stable isotope work in the Copper River system. This was rated highly by the peer reviewers because there is a likelihood of a spill due to the pipeline and there is no information on this area. This requires a matching grant to fund: Kline has until the end of FY99 to acquire a match. Regarding the proposal to archive EVOS intertidal monitoring samples, more information and matching funds have been requested from the proposer. The pink salmon monitoring proposal by Mark Willette will allow a determination of ocean mortality vs. in-Sound mortality. Award is pending support from ADF&G. There have been four proposals for science planning workshops, one from the original SEA group to revisit the SEA plan, one from ACOPS, one from a Bristol Bay group, and one from a Cook Inlet group. Only the SEA group proposal included all groups: users, researchers, and managers. Thomas suggested that there be a workshop to educate people on how to put these types of projects together.

Thomas anticipates that projects will come out of Cordova District Fishermen United workshop on small spills. There is $35,000 set aside for this.

Break

Meeting recessed at 11:15 a.m. for 15 minutes.

Continuation of Director's Report

Technology Coordinator Position: There are seven applicants as of this date. Thomas suggested working with the Executive Committee to review the applicants.

Action item

Thomas is directed to interview the applicants and the Executive Committee will meet in a few weeks to consider applicants. A list of the applicants will be distributed to all Board members as a personal/confidential communication.

Grant Policy Manual

Indirect costs: Thomas requested some flexibility in the 20% indirect costs ceiling for grants. One possibility is to accept the federal audited rates for organizations. Thomas believes that the 20% ceiling mandated in the legislation for OSRI is for administrative costs only, and that the Board can change it for grants. Moore says this is how NOAA operates. The University of Alaska Fairbanks has a 25% indirect cost rate negotiated with the EVOS Trustees.

MOTION

Motion by Kopchak to allow contracts issued by OSRI to use the federal audited indirect cost rates, but to encourage negotiating lower indirect costs. No second. Discussion. Issue tabled until after lunch pending a revised motion.

Kopchak requested that there be links to documents referenced in the Grant Manual on the web page. Moore pointed out that the Grant Policy Manual is a living document and will continually change. It can be approved as stands with the understanding it will change over time. There are two major issues still to resolve: indirect costs and the payback policy. The approval vote will be postponed until these two issues are discussed after lunch.

Business Plan

The current version incorporates some of Dietrick's comments and integrates the newest information from the development of the SEA circulation model. It was decided to give the Board more time to review and comment on the current version. Comments are requested by October 1.

Action Item

OSRI staff will prepare a revised business plan after receipt of comments. It will be sent to all Board members prior to scheduling an executive committee meeting. The Board will then be polled by Bird to find out if the Board is comfortable with delegating approval to the Executive Committee. The draft plan will be put on the web page with an invitation for public comments until Oct. 1. (*Editor's note*: See action at end of meeting scheduling a teleconference meeting of the full Board on October 6 to consider several items, including the Business Plan; so that comments received can be incorporated into a revised document and sent to Board members prior to this meeting date, public and individual Board comments on the Business Plan are requested by Sept. 15.)

Lunch Break   Adjourned at 12:20 p.m., reconvened at 1:30 p.m.

Scientific and Technical Committee Report   Goering said that the proposals received are reviewed to see if they fully address the relevant BAA and, if so, they are sent out for peer review as established in the committee guidelines.

Circulation model   The PWSSC proposal for PWS was the only one that filled all the requirements of the BAA, but the original request was too high. A scaled down proposal in the area of $300,000/year has been received by OSRI, and it is being sent out for peer review. Upon receipt of the peer reviews the committee will meet and give a go or no go recommendation to the Board. This may require a teleconference meeting with the Board in October. Moore requested a copy of the project summary.

Action item   Prior to the teleconference the OSRI staff will distribute copies of the project summary, the peer reviewer's comments and the recommendation of the committee to the Board.

Goering said that the committee intends to sponsor a workshop in the Cook Inlet area to help interested parties develop a Nowcast/Forecast proposal for the area that more adequately addresses the requirements of the BAA. The committee received some proposals for observational programs in the Gulf of Alaska but have decided not to get involved with these efforts at present. There are other programs coming on line that may fund these, and OSRI could work on linking all of this together in the future.

1999 Annual   Thomas pointed out that the Nowcast/Forecast circu-

Work Plan

lation model program money is coming 1/2 from technology funds and 1/2 from ecology funds. It is a 5 year program with $1.5 million of dedicated funding. It is all dedicated in FY99 but only $300,000 will be spent in FY99. This is $150,000 of the $400,000 available in each of the technology and ecology programs per annum. The business plan includes BAA's about remote sensing and dispersant effectiveness with funding up to $200,000 to see what is out there: these were relevant issues identified at the ice hazards workshop. BAA's on oil toxicity and monitoring have been issued and proposals are expected.

Treadwell pointed out that OSRI is hoping to match the $300,000 for the nowcast/forecast project twice, by EVOS, and by industry. Thomas said that SERVS has already promised a commitment of vessels and personnel. Wright commented that EVOS is not comfortable with supporting multi-year projects at this time but has indicated some support for this project probably at less than $300,000.

Dietrick expressed concern that enough money be set aside for technology research and development. Thomas said that a BAA was issued for this but few proposals were received. Treadwell says private companies are working on oil/water separators and fast water flow oil booms and he will submit the names.

Thomas quickly ran through list of projects in the FY99 business plan, commenting that the staff is working hard to achieve a 40/40/20 split, between technology, ecology and education and has been largely successful. He quickly commented on line items in the FY99 business plan. *Sensitive area mapping:* $100,000 in commitments have been made but $50,000 more is needed according to Mutter. *Video:* If the first one is successful more will be considered that are more hardware oriented but for the general public. *Annual report:* A 10 min. discussion ensued on whether $15,000 was enough to produce an annual report, on whether the report is an administrative or a program expense, and on what format should be used.

Treadwell commented that the PWSSC Board would like to see $100,000 made available to help make the SEA plan modeling efforts become practical, and also transferrable to other areas. NOAA, NMFS, the North Pacific Council, and the Atlantic Council have all expressed interest in this and would like to see a guide produced detailing how to make a SEA type process happen.

MOTION

Motion by Treadwell to adopt the FY99 Annual Work Plan. Second with amendment by Dietrick to change the ratio of 50/50 funding for the nowcast/forecast model from technology/ecology to 66/33, but agreed to revisit the issue next year. Amendment changed to move the $50,000 of unfunded ecology money from FY98 to technology in FY99.
Motion as amended passed unanimously.

Break

Meeting recessed at 3:25 p.m. for 15 minutes.

Treasurer's Report

Kopchak presented the new budget report format. Vidrine suggested adding a one page summary at the front. Treadwell suggested adding income also. Dietrick would like to have a footnote showing the amount encumbered for a project over future years as well as the present year. Treadwell summarized the Morgan Stanley investment statement, noting that the return is better than CD's would have been, but would be better still if OSRI could invest longer term.

Action Item

Kopchak will work with Oswalt to prepare a summary report to the Board prior to the fourth quarter. In future the Board will receive a summary report: any member that wishes can request the entire report.

MOTION

Motion to approve the 3rd quarter report and the FY99 budget, seconded by Leland.
Motion passed unanimously.

Grant Policy Manual

(cont. from morning)

MOTION	Motion by Kopchak, seconded by Dietrick to establish a policy authorizing the Executive Director authority to consider indirect cost rate higher than that authorized in the grant manual based on the project's merit in response to the BAA or RFP; however, no indirect cost rates may exceed the federally negotiated rate. **Motion passed** unanimously.
MOTION	Motion by Kopchak that the staff develop substitute language for the intellectual property and program income sections to be developed and presented to the Board within 90 days. **Motion passed.**
MOTION	Motion by Mutter, seconded by Kopchak to change the language at the top of page 48 by deleting "maximum" and "The OSRI Advisory Board approved a similar restriction." **Motion passed** unanimously.
MOTION	Motion by Treadwell, seconded by Vidrine to add the sentence, "This policy is currently under review by the Board, and grants and contracts may be subject to a revised policy to be approved during FY99." **Motion passed** unanimously.
MOTION	Motion by Leland, seconded by Dietrick, to approve the Grant Policy Manual with amendments. **Motion passed** unanimously.
Bylaws Review	Motion by Treadwell that Kopchak is in a conflict of interest if he votes for the budget because his wife is employed by a business that has been awarded an OSRI contract. No second. Chair Moore declared that this means the Board does not find a conflict of interest.
Discussion	A discussion ensued about the conflict of interest policy, with differing opinions about whether the state policy applies to all Board members. It was pointed out that there may be a difference between small grants and large grants, as grants under $100,000 are not approved individually by the Board. It was agreed that

more work needs to be done on this issue before it can be resolved.

Action Item   The staff is directed to do more homework and to provide a draft conflict of interest policy for discussion by the Board members at the next teleconference meeting.

Grant   The issue of the necessity of all grants being nationally
Advertising   competitive was revisited. Mutter and Moore said that
(cont. from   upon review of the actual law, they think the
morning)   Brookings letter expresses the correct opinion which is that not all grants have to be nationally advertised. Treadwell reiterated his opinion that an annual BAA be issued that is a general policy statement, inviting responses to any programs that OSRI is pursuing. The idea of a dollar limit on what grants would be advertised nationally was discussed, and what sort of mailing list RFP's should go out to. Leland said that RCAC keeps a database of interested parties for this purpose, but Mutter expressed concern about the administrative costs of keeping such a database up to date. It was pointed out that the Board has already adopted a policy (page 35 of the Grant Policy Manual), but it was also felt that more explicit language might be necessary.

Action Item   Staff was directed to review this issue further, and if they see a need to modify the policy, to come back to the Board with written recommendations.

Next Meeting   There will be a teleconference meeting Tuesday, Octo-
Dates   ber 6, to discuss the business plan, the nowcast/forecast proposal, and the conflict of interest policy statement.
There will be a full Board meeting the week of April 12 in Seward, Alaska.

Public Comments None.

Board Members   Mutter brought up two topics of potential concern and
Closing   possible funding. The first is what kind of toxic effects dispersants might have on the biota. The other concerns derelict ships, how many are out there, and what

are the pollution consequences of them breaking up, as in the recent incident on St. Matthew's Island.

Tuxhorn pointed out that there is a spill of opportunity coming up in September, with the Sound-wide drill on Sept. 18. He also said a learning center is opening at the Alaska SeaLife Center on Sept. 28 that the OSRI might want to collaborate with.

Adjournment   **Motion** by Kopchak, seconded by Leland to adjourn. Meeting adjourned at 5:15 p.m.

These minutes were reviewed and approved by the Advisory Board on November 24, 1998. Doug Mutter, Secretary

D

PWSSC/OSRI Proposal Review Form

Prince William Sound Oil Spill Recovery Institute
P.O. Box 705 - Cordova, AK 99574 - (907) 424-5800; fax 424-5820
e-mail: frontdes@pwssc.gen.ak.us

****
PROPOSAL REVIEW FORM

****

Guide for Reviewers

Thank you for your expertise and time in reviewing this proposal. We need an impartial and professional evaluation. If you find you have a conflict of interest, please notify us and return this proposal.

You may range as widely in your comments as you wish; however, please strive to make your comments both constructive and civil, even for unfavorable assessments. You are encouraged to point out the proposals strong points as well as the weak ones. You need not be concerned about the proposal format.

Please either type your comments on separate sheets of paper and attach this form, or e-mail the written comments to bird@pwssc.gen.ak.us. We request that all proposals be treated as the intellectual property of the principal investigators or their employers. Their confidentiality must be respected during the review process.

Please complete the rating summary below: If you <u>do not</u> wish to remain anonymous, sign the review. Return your review and the proposal directly to me. Also, retain a copy for your own files to guard against loss in the mail. Your assistance is greatly appreciated. Thank you!

Evaluation criteria:

(1) Is the proposal clearly written and understandable?

(2) Are there clear objectives for the project?

(3) Is there a clear outline of organization for its implementation?

(4) Does the project propose multi-disciplinary or multi-organizational coordination? Or, does it include partnerships with industry, agencies and others to accomplish its goals? Are there entities missing which should be involved in this project?

(5) Does the proposed work have a low, moderate or high chance of contributing new information, developing a new method or providing an unique service to its field?

(6) If successful, would the proposal make an unique, major, moderate, minor or insignificant contribution?

(7) Do the investigator(s) identify who the major users of the results will be?

(8) Do the investigator(s) have the expertise to implement this project?

* * * * * * * * * * * * * *

Rating summary (check one):

Fund ____

Fund after minor revisions ____

Reconsider after major revision ____

Reject ____

Better suited for other funding ____

Reviewer's signature (optional, but encouraged):

E

Membership of OSRI/PWSSC Advisory Groups

OSRI Advisory Board Members:

Virginia Adams
John Calder, National Oceanic and Atmospheric Administration
Capt. Jack Davin, U.S. Coast Guard
Gail Evanoff
Mark Fink, Alaska Department of Fish & Game
Carol Fries, Alaska Department of Natural Resources
John Goering, Ph.D., University of Alaska, Fairbanks
R.J. Kopchak
Marilyn Leland, Prince William Sound Regional Citizens' Advisory Council
Doug Lentsch, Cook Inlet Spill Prevention & Response, Inc.
Doug Mutter, U.S. Department of the Interior
Walter B. Parker, *Chair,* Prince William Sound Science Center
Leslie Pearson, Alaska Department of Environmental Conservation
Edmond Paul Thompson, BP Oil Shipping Company
Glenn Ujioka

OSRI Committees – as of March 2002

Executive Committee

John Calder, Ph.D., Chair, *(Department of Commerce representative is designated chair)*

Doug Mutter, Vice Chair, Fall 2003
R.J. Kopchak, Treasurer, Fall 2003
Marilyn Leland, Secretary, Fall 2003
Ed Thompson, Fall 2002
Brad Hahn, Fall 2002

Finance Committee – appointed November 1999

R.J. Kopchak, Chair/Treasurer
Carol Fries
Ed Thompson
Mead Treadwell (nonvoting)

Scientific and Technical Committee

John Goering, Ph.D., Chair, Professor Emeritus, Institute of Marine Science, School of Fisheries and Ocean Sciences, University of Alaska Fairbanks
Douglas M. Eggers, Ph.D., Chief Scientist, Division of Commercial Fisheries, Alaska Department of Fish & Game
Raymond Jakubczak, Ph.D., BP Exploration Alaska Pipeline Unit, BP Pipelines Alaska
Alan J. Mearns, Ph.D., BioAssessment Division, National Oceanic & Atmospheric Administration
Stanley (Jeep) Rice, Ph.D., Alaska Fisheries Science Center, National Marine Fisheries Service, Auke Bay Lab
Thomas C. Royer, Ph.D., Center for Coastal Physical Oceanography, Old Dominion University

FY03 Work Plan Committee – appointed February 2002

Doug Lentsch
Mark Fink
Brad Hahn
Ed Thompson
Marilyn Leland

PWSSC Board of Directors

Executive Committee

Walter B. Parker, Chair, Commissioner, U.S. Arctic Research Commission, Anchorage, 1997-2003

Charles R. Parker, 1st Vice Chair, Executive Director, Mat-Su Resource & Conservation Development, Inc., Wasilla, 1993-2002

Chris Blackburn, 2nd Vice Chair, Retired Director, Alaska Groundfish Data Bank, Kodiak, 1997-2003

Meera Kohler, Treasurer, President & CEO, Alaska Village Electric Coop., Inc., Anchorage, 2000-2003

Charles Meacham, Secretary, President, Capital Consulting, Juneau, 1997-2003

Mead Treadwell, Member-at-large, Managing Director, Institute of the North, Anchorage, 1994 –2003

Members

John Allen, Chair, PWS Regional Citizens' Advisory Council, Valdez, Alaska, 2001-2004

Ed Backus, Director of Community and Salmon Programs, Ecotrust, Charleston, Oregon, 2001-2004

Gail Evanoff, Chenega Bay, Alaska, 1996-2004

Simon Lisiecki, Senior Vice President, BP, Anchorage, 2001-2004

Ole Mathisen, Ph.D., Former Dean, Juneau School of Fisheries & Ocean Sciences, University of Alaska, Juneau, 1995-2004

Trevor McCabe, Executive Director, At-Sea Processors, Anchorage, 2000-2003

Stu Nozette, Washington, D.C., 2001-2004

Steven D. Taylor, Ph.D., Retired Manager, Environmental & Regulatory Affairs, BP Exploration Inc., Anchorage, 1993-2002

Gary L. Thomas, Ph.D., President, Prince William Sound Science Center, Cordova, Ex officio member

David B. Witherell, Fisheries Management Biologist, North Pacific Fishery Management Council, Anchorage, 2000-2003

Edward Zeine, Former Mayor, City of Cordova, 2001-2004

Board Members Emeriti

John P. Harville, Ph.D., Interim Director, Prince William Sound Science Center, 1989-90, Former Director, Pacific Marine Fisheries Commission, Portland

Calvin Lensink, Retired, U.S. Fish & Wildlife Service, Anchorage

Nolan Watson, Senior Vice President, McClellan & Copenhagen, Inc., Seattle

F

Broad Area Announcements

Broad Area Announcement
for Applied Technology Program of the
Prince William Sound
Oil Spill Recovery Institute

Deadline for receipt of applications: December 31, 1999
Date this BAA was posted: October 6, 1999

The OSRI announces a competition for <u>Computer Simulation of the Spatial-Temporal Distribution and Impacts of Dispersed and Non-Dispersed Oil Spills</u>

The Oil Spill Recovery Institute is accepting proposals for technologies which demonstrate three dimensional trajectories as well as the resulting physical and biological environmental impacts of dispersed and non-dispersed oil in arctic and subarctic marine environments. Proposals which focus on Alaskan oil transportation routes and utilize realistic models will be given preference.

The total FY99 budget for this program area is $200,000 and OSRI anticipates funding one or more projects from this total amount. The duration of the grant awards will be for one year with an option to renew.

Request for Proposals
Predictive Ecology Program of the
Prince William Sound
Oil Spill Recovery Institute

Request for Proposals – Biological Monitoring (Intertidal Resources at Risk to Oil Spills)
Deadline for receipt of applications: January 31, 2000
Date this BAA was posted: December 28, 1999

The Oil Spill Recovery Institute (OSRI) is seeking proposals to conduct surveys of unique, dominant and/or previously undescribed, intertidal resources at risk to oil spills in Prince William Sound (PWS) and the Copper River Delta (CRD). OSRI anticipates surveys staging out of Cordova, Alaska.

Professional services sought through this RFP include the survey prelogistics (design, equipment preparation, etc.), data acquisition, data analysis (including preparing a GIS of the data), and reporting of results. Data acquisition and analysis will require coordination with the OSRI personnel who are responsible for maintaining the database. The maximum amount of funding for this contract is $75,000. The selected contractor will provide and operate all scientific equipment. All sampling shall be conducted synoptic with GPS measurements, ensuring the acquisition of data that is both temporally and geographically encoded. Contractor will provide and operate all necessary computer equipment and software to acquire and process data. Contractor will cooperate with other teams of researchers and technologists that OSRI might employ to advance the ability to monitor intertidal or adjacent areas that are at risk to oil spills.

The selected contractor shall prepare a preliminary report detailing the survey methods, the analysis of measurements and the anticipated results one month after the end of the field season. The contractor is responsible for providing an electronic copy of the 2d survey in ARCINFO file format with the final report. The final report and data file is due on December 31, 2000.

The survey design will utilize all available aerial, vessel and historical information on the distribution of resource or its habitat to reduce size of the survey to a reasonable area. Where large patches of organisms are found, repeated surveys will be conducted to estimate precision. On each survey as many samples will be conducted as possible to satisfy statistical requirements. Quasi-continuous sampling with optical measurement is preferred over discrete sampling and non-quantitative sampling is discouraged. Prospective contractors interested in submitting proposals should direct technical questions and project proposals to the OSRI staff (Gary Thomas, Walter Cox or Nancy Bird) at P.O. Box 705, Cordova AK 99574, (907) 424-5800. Proposals should contain the name and address of the firm; name, address, phone and fax numbers of the contact person for the proposal; a comprehensive description of the equipment and procedures to be used; as well as a brief description of the qualification of the firm and key personnel. Experience of key personnel is critical. The deadline for proposals is January 31, 2000 and proposals must be received by the deadline to be considered. Contractors wishing to submit a proposal are advised that qualification and capabilities will be considered in the evaluation process. Specific criteria that will be used to evaluate proposals include:

1. The capabilities, related experience, facilities, equipment, techniques and methods of the proposing organization.

2. The capabilities, qualifications and experience of the proposing organization's key personnel involved in actively performing the data acquisition, analysis and reporting.

3. The organization's record of past performance with similar types of projects. Applicants should include references to aid in evaluation of performance criteria.

4. The organization's estimated cost to perform the required professional services.

Intertidal survey schedule:
December 28, 1999 Solicitations issued
January 31, 2000 Final Day for Submission of Proposals
February 15, 2000 Award Contract
Preliminary report due one month after end of field season
November 1, 2000 Draft final report and data file due
January 31, 2000 Final report as a manuscript submitted for publication
Application Process

Broad Area Announcement
for Applied Technology Program of the
Prince William Sound
Oil Spill Recovery Institute

Deadline for receipt of applications: May 15, 2002
Date this BAA was posted: March 13, 2002

The OSRI announces a competition for Applied Spill Response Planning

The Oil Spill Recovery Institute is accepting proposals for grants that focus on application of new and developing spill response and response planning techniques for the Kachemak Bay Zone of the Cook Inlet Subarea. Proposals which build upon the OSRI geographic response plans workshop proceedings entitled "Geographic Response Plans for Alaska" (available through OSRI), which outlines techniques new to Alaska for designing spill response plans for the protection of priority sensitive areas along the coast, will be given preference. Local spill response organizations are encouraged to apply. The total FY02 budget for this program area is $25,000. The duration of the grant awards will be for one year.

G

OSRI: Final Report (1992-1995)

Submitted to National Oceanic and Atmospheric Administration
March 1996
Summary

The Oil Pollution Act of 1990 authorized establishment of the Prince William Sound Oil Spill Recovery Institute to serve two primary functions:

1. to identify and develop the best available techniques, equipment and materials for dealing with oil spills in the Arctic and Subarctic marine environment; and

2. to complement damage assessment efforts of state and federal agencies and determine, document, assess, and understand the long-range effects of the Exxon Valdez oil spill on the natural resources of Prince William Sound and its adjacent waters.

The 19-member Advisory Board of the Oil Spill Recovery Institute (OSRI) was appointed by the Secretary of Commerce in 1992 and the first Board meeting was held in October 1992. This Advisory Board included six Federal agency representatives, four representatives from State of Alaska agencies, four citizens from Alaskan communities impacted by the Exxon Valdez oil spill, and three representatives from Native communities in the oil spill impacted region. Additionally, non-voting representatives served on the Board from the University of Alaska Fairbanks and from the Prince William Sound Science and Technology Institute (aka Prince William Sound Science Center or PWSSC).

Seven meetings of the Advisory Board and its Executive Committee were held between 1992 and 1995. A six-member Scientific and Technical

Committee was appointed in 1993 and met several times to review research plans and proposals.

The PWSSC, a non-profit research and education organization based in Cordova, Alaska, was awarded a Federal grant totaling $480,000 to administer the OSRI and share its professional scientific and administrative staff. The PWSSC staff provide scientific expertise in the areas of oceanography, geographic information systems, fisheries ecology, remote sensing, advanced visualization and education.

The PWSSC also provided administrative staff for the Alaska Hazardous Substance Spill Technology Review Council (HSSTRC) which was created to assist state agencies in the identification of containment and clean-up products and establishing optimum oil spill prevention and response techniques.

These two state and federal organizations (HSSTRC and OSRI) collaborated in the preparation of a document identifying the most critical research necessary for oil spill prevention and response in Arctic and Subarctic waters. The Oil Pollution Research and Technology Plan for the Arctic and Subarctic was drafted in 1994. In 1995, it was submitted for peer review and was critiques by both the OSRI and HSSTRC Advisory Boards. A revised document was published in December of 1995.

Accomplishments, 1992 - 1995
in relation to the OSRI's two stated missions:

1. *To identify and develop the best available techniques, equipment and materials for dealing with oil spills in the Arctic and Subarctic marine environment.*

- Published the Oil Pollution Research and Technology Plan for the Arctic and Subarctic. This plan established research priorities to attain the most efficient and effective technologies and methods for preventing and responding to oil spills in the Arctic and Subarctic.
- Assisted the Alaska Department of Environmental Conservation in soliciting and reviewing proposals in 1994 and 1995 for oil pollution research projects to be funded by a grant from the Alaska Legislature. Over 100 proposals were received and approximately six were awarded funding.
- Established a library of reports, books and other resources on the subject of oil pollution prevention and response techniques in cold waters.
- Assisted in the distribution of a 10-minute videotape on the latest results from an in-situ burn of oil off the Newfoundland coast.

2. *To complement damage assessment efforts of state and federal agencies and determine, document, assess, and understand the long-range effects of the Exxon Valdez oil spill on the natural resources of Prince William Sound and its adjacent waters.*

• OSRI and PWSSC staff were leaders in the development of an ecosystem research program designed to identify the processes affecting the biomass, production and behavior of marine populations in the Prince William Sound and North Gulf of Alaska ecosystems. Called the Sound Ecosystem Assessment (SEA), the program is a multi- disciplinary study being carried out by scientists at the PWSSC, state and federal agencies and the University of Alaska Fairbanks. The results from this long-term (5-8 years) study will be used by the Exxon Valdez Oil Spill Trustee Council to determine the long-range effects of the oil spill. The EVOS Trustee Council provides the majority of funding for the SEA project, but the OSRI budget provided geographic information system personnel who contributed to the program. In turn, the data collected by the SEA projects are increasing the database for the OSRI geographic information system of the region.

• Education staff of the PWSSC first published the Alaska Oil Spill Curriculum in 1990, prior to the establishment of the OSRI. This curriculum offers projects on oil pollution, prevention and conservation for ages pre-school through high school. It was widely distributed throughout Alaska in 1991 and also to school districts in other coastal areas of the United States; requests for the curriculum have come from teachers in New Zealand, Norway and several other foreign countries. In 1995, a revision of the curriculum was completed with support from the OSRI budget.

H

OSRI Journal and Workshop Publications

Alaska Clean Sea. 2001. International Oil & Ice Workshop 2000: Oil Spill Preparedness for Cold Climates (CD). Proceedings of the Conference, April 5-7, 2000, Anchorage & Prudhoe Bay, Alaska. Alaska Clean Sea.

Bang, I., Mooers, C.N.K. 2002. The influence of several factors controlling the interactions between Prince William Sound, Alaska, and the Northern Gulf of Alaska. *Journal of Physical Oceanography*. In press.

Ben-David, M., T.M. Williams, and O.A. Ormseth. 2000. Effects of oiling on exercise physiology and diving behavior of river otters: a captive study. *Canadian Journal of Zoology*. 78:1380-1390

Ben-David, M., G. Blundell, and J. Blake. Post-release survival of river otters: effects of exposure to crude oil and captivity. Submitted

Bishop, M.A., and S.P. Green. 2001. Predation on Pacific herring *(Clupea pallasi)* spawn by birds in Prince William Sound, Alaska. *Fisheries Oceanography* 10(1):149-158.

Braddock, J.F., and K.A. Gannon. 2001. Petroleum hydrocarbon degrading microbial communities in Beaufort Sea sediments. Minerals Management Service Information Transfer Meeting, Anchorage, AK, April 2001.

Clarke, E.D., L.B. Spear, M.L. McCracken, D.C. Borchers, F.F.C. Marues, S.T. Buckland, and D.G. Ainley. Application of generalized additive models to estimate size of seabird populations and temporal trend from survey data collected at sea. *Journal of Applied Ecology*. In press.

Chesceri, E.J., J.H. Grabowski, S. Powers, C.H. Peterson, C.S. Martens, and M. Bishop. 2001. Carbon and nitrogen isotopic tracing of organic matter flow in two tidal estuaries of southcentral Alaska. Presentation by E.J. Chesceri: North American Benthological Society Meeting, LaCrosse, WI, June, 2001.

Cook Inlet Regional Citizen's Advisory Council. 1999. Forum Proceedings: Safety of Navigation in Cook Inlet, September 9 & 10, 1999. Cook Inlet Regional Citizen's Advisory Council, Kenai, AK.

Ferry, Leann, Lisa Ka'aihue, and Linda Robinson. 2002. Peer Listener Training Video. Prince William Sound Oil Spill Recovery Institute, Cordova, Alaska.

Gannon, K.A., and J.F. Braddock. 2001 Petroleum hydrocarbon degrading microbial communities in Beaufort Sea sediments. Oral presentation at the American Society for Microbiology-Alaska Chapter 17th Annual Meeting, Anchorage, AK, April 2001.

Hartwell, Kevin. 2001. Celebrating Alaska's Shorebirds (video). Prince William Sound Oil Spill Recovery Institute, Cordova, Alaska.

Johnson, Mark A., and Stephen R. Okkonen (ed). 2000. Proceedings of the Cook Inlet Oceanography Workshop, 9 November 1999, Kenai, Alaska. Institute of Marine Science, University of Alaska, Fairbanks, Alaska.

Kirsch, Jay, and G.L. Thomas, 2000. Acoustic estimation of spring plankton densities in Prince William Sound. *Fisheries Research.* 47(2000):245-260

LaFornaise, John. 1999. Sound Science (video). Prince William Sound Oil Spill Recovery Institute, Cordova, Alaska.

Mooers, C.N.K., I. Bang, and S.L. Vaughan. Experience with the Prince William Sound Nowcast/Forecast System. AMS Preprint Volume.

Mooers, et al. A Nowcast/Forecast System for Prince William Sound. In preparation .

National Oceanic & Atmospheric Administration, Office of Response and Restoration. 2000. Sensitivity of Coastal Environmental and Wildlife to Spilled Oil: Prince William Sound, Alaska. National Oceanic and Atmospheric Administration, Silver Spring, Maryland.

Oedekoven, Cornelia, David G. Ainley, and Larry B. Spear. Variable responses of seabirds to change in marine climate: California Current, 1985-1994. *Marine Ecology Press Series.* In press.

Owens, Edward H., Laurence B. Solsberg, Mark R. West, and Maureen McGrath. 1998. Field Guide for Oil Spill Response in Arctic Waters. Environment Canada, Yellowknife, NT Canada.

Powers, S.P., and H.S. Lenihan. 2002. Physical biological coupling and the conservation of marine species. *Journal of Sea Research.* In review.

Powers. S.P., and J.N. Kittinger. 2002. Hydrodynamic mediation of predator-prey interactions: differential patterns of prey susceptibility and predator success explained by variation in water flow. *Journal of the Exploration of Marine Biology and Ecology.* In press.

Powers, S.P., and M.A. Bishop. 2001. Ecology of the Copper River Delta, Alaska: potential role of avian predators in determining macrobenthic community structure. Presentation by S.P. Powers: Benthic Ecology meeting, University of New Hampshire, Durham, March, 2001.

Powers, S.P., M.A. Bishop, J.H. Grabowski, and C.H. Peterson. 2002. Intertidal benthic resources of the Copper River Delta, Alaska, USA. *Journal of Sea Research* 47(1):13-23.

Powers, S.P., M.A. Bishop, J.H. Grabowski, and L. Manning. Growth, recruitment, and population dynamics of *Mya arenaria*, an invasive species on the Copper River Delta. *Journal of Shellfish Research.* In preparation.

Prince William Sound Oil Spill Recovery Institute. 1998. Proceedings of A Symposium on Practical Ice Observation in Cook Inlet and Prince William Sound, January 6-7, 1998, University of Alaska, Anchorage. Prince William Sound Oil Spill Recovery Institute, Cordova, AK.

Prince William Sound Regional Citizens' Advisory Council. 2001. Peer Listener Training Video. Prince William Sound Regional Citizens' Advisory Council, Valdez, AK.

Research Planning, Inc. 2001. Coastal Resources Inventory and Environmental Sensitivity Maps: Aleutians West Coastal Resources Service Area, Alaska. Research Planning, Inc., Columbia, South Carolina.

Robertson, Tim L. (ed.). 1999. Geographic Response Plans for Alaska Workshop Report: November 17, 1998. Prince William Sound Oil Spill Recovery Institute, Cordova, Alaska.

Robertson, Tim L., and Elise DeCola (ed.) 2001. Final Proceedings of the Prince William Sound Meteorological Workshop, December 12th to 14th, 2000, Anchorage, Alaska. Prince William Sound Oil Spill Recovery Institute, Cordova, Alaska.

Thomas, G.L., and Walter Cox. 2000. A Nowcast/Forecast information system for Prince William Sound. Proceedings of the 23rd Arctic and Marine Oil Spill Program (AMOP) Technical Seminar, Environment Canada, Ottawa, Ontario, Canada. Vol. 1:247-255.

Thomas, G.L., and Jay Kirsch (Guest Editors). 2000. Special Issue: Recent advances and applications of acoustics to fisheries research. *Fisheries Research.* 47(2000):107-302.

Thomas, G.L., and Jay Kirsch. 2000. Nekton and plankton acoustics: An overview. *Fisheries Research.* 47(2000):107-113

Thomas, G.L., J. Kirsch, and R.E. Thorne. 2002. Ex situ target strength measurement of Pacific herring and Pacific sand lance. *North American Journal of Fisheries Management.* In press.

Thomas, G.L., and R.E. Thorne. 2001. Abundance and distribution of walleye pollock in Prince William Sound. Electronic Proc. ACOUSTGEAR 200. International Symposium on Advanced Techniques of Sampling Gear and Acoustical Surveys for Estimation of Fish Abundance and Behavior, Hakodate, Japan, October 20-21, 2000.

Thomas, G.L., and R.E. Thorne. 2001. Night-time predation by Stellar Sea Lions. *Nature* 411:1013.

Thomas, G.L., R.E. Thorne, and W.R. Bechtol. 2001. Developing an effective monitoring program for pollock in Prince William Sound, Alaska. Electronic Proc. ACOUSTGEAR 200. International Symposium on Advanced Techniques of Sampling Gear and Acoustical Surveys for Estimation of Fish Abundance and Behavior, Hakodate, Japan, October 20-21, 2000.

Thorne, R.E., and G.L. Thomas. 2001. Biological considerations for scaling ecosystem research in Prince William Sound, Alaska. Electronic Proc. ACOUSTGEAR 200. International Symposium on Advanced Techniques of Sampling Gear and Acoustical Surveys for Estimation of Fish Abundance and Behavior, Hakodate, Japan, October 20-21, 2000.

Thorne, R.E., and G.L. Thomas. 2001. Monitoring the juvenile pink salmon food supply and predators in Prince William Sound. In: R. Beamish, Y. Ishida, V. Karpenko, P. Livingston, and K. Myers (eds.), Workshop on factors affecting production of juvenile salmon: comparative studies on juvenile salmon ecology between the east and west North Pacific Ocean, Technical Report 2, North Pacific Anadromous Fish Commission, Vancouver, B.C. pp 42-44.

Trudel, Ken (ed.). 1998. Proceedings of the Conference, "Dispersant Use in Alaska: A Technical Update," Anchorage, Alaska, March 18 & 19, Prince William Sound Oil Spill Recovery Institute, Cordova, Alaska.

Trudel, B.K. 1998. Environmental risks and trade-offs in Prince William Sound. In: B.K. Trudel, (ed.), Proceedings of the Conference, Dispersant Use in Alaska: A Technical Update, Anchorage, Alaska, March 18 & 19, 1998, Prince William Sound Oil Spill Recovery Institute, Cordova, Alaska.

Trudel, B.K. 1998. Monitoring the effectiveness and effects of dispersant operations. In: B.K. Trudel, (ed.), Proceedings of the Conference, Dispersant Use in Alaska: A Technical Update, Anchorage, Alaska, March 18 & 19, 1998, Prince William Sound Oil Spill Recovery Institute, Cordova, Alaska.

Trudel, K., P. Armato, D. Maguire, S. Hillman, B. Hahn, L. Pearson, R. Morris, and D. Rome. 1999. Dispersant use in Alaska: an update. In: Proceedings of 1999 International Oil Spill Conference, Seattle, Washington, pp 807-812.

Vaughan, S.L., and W. Cox. 1999. A Nowcast/Forecast system for Prince William Sound (abstract). In: Proceedings of the Cook Inlet Oceanography Workshop, 9 November 1999, Kenai, Alaska. 2000. M.A. Johnson and S.R. Okkonen (eds.). Institute of Marine Science, University of Alaska, Fairbanks, Alaska.

Vaughan, S.L., and S.M. Gay. 2000. Physical processes influencing biological components of Prince William Sound, Alaska (abstract). Presentation at the AGU 2000 Ocean Sciences Meeting, Jan. 2000, San Antonio, Texas.

Vaughan, S.L. 2001. A Nowcast/Forecast system for Prince William Sound: Observational oceanography (abstract). Presented at the Fifth International Marine Environmental Modeling Seminar (IMEMS 2001), Oct. 9-11, 2001.

Vaughan, S.L., and S.M. Gay. 2001. Seasonal hydrography and tidal currents of bays and fjords in Prince William Sound, Alaska. *Journal of Fisheries Oceanography.* In press.

Vaughan, S.L., C.N.K. Mooers, and S.M. Gay. 2001. Physical variability in Prince William Sound during the SEA study. *Journal of Fisheries Oceanography.* In press.

Wang, J., M. Jin, V. Patrick, J. Allen, C.N.K. Mooers, D.L. Eslinger, and T. Cooney. 2000. Numerical simulation of the seasonal circulation patterns and thermohaline structures of PWS, Alaska. *Fisheries Oceanography:* SEA Synthesis Vol. (conditionally accepted).

I

Acronyms and Initialisms

ADF&G   Alaska Department of Fish and Game
API   American Petroleum Institute
ARCUS   Arctic Research Consortium of the United States
ATOM   Alyeska Tactical Oil Spill Model

BAA   Broad Area Announcement
BLM   Bureau of Land Management

CGAA96   Coast Guard Authorization Act of 1996
CI   Cook Inlet
CMI   Coastal Marine Institute
CoOP   Coastal Ocean Processes program (NSF, NOAA, ONR)

DOD   Department of Defense

EPPR   Emergency Prevention, Preparedness, and Response
ESI   environmental sensitivity maps
EVOSTC   *Exxon Valdez* Oil Spill Trustee Council

FOSC   federal on-scene coordinator
FTE   Full-Time Employee
FY   fiscal year

GEM   Gulf (of Alaska) Ecosystem Monitoring
GIS   geographic information system

GLOBEC   Global Ocean Ecosystem Dynamics program
GPM   Grant Policy Manual
GPS   Global Positioning System
GRD   Graphical Resource Database

HAZMAT   HAZardous MATerials
HE   Hinchinbrook Entrance
HSVA   Hamburgische Schiff Versuchs Anstalt (German ice model)

IARC   International Arctic Research Center
IMS   Institute of Marine Sciences

LIDAR   LIght Detection And Ranging
LTEMP   Long-Term Environmental Monitoring program

MORICE   Mechanical Oil Recovery in Ice Infested Waters
MOU   memorandum of understanding
MS   Montague Strait

NC/FC   Nowcast/Forecast Model
NMFS   National Marine Fisheries Service (NOAA)
NOAA   National Oceanic and Atmospheric Administration
NOS   National Ocean Service
NPRB   North Pacific Research Board
NRC   National Research Council
NRDA   Natural Resource Damage Assessment
NSF   National Science Foundation

OAR   Office of Oceanic and Atmospheric Research
ONR   Office of Naval Research
OPA 90   Oil Pollution Act of 1990
OR&R   Office of Response and Restoration
OSCAR   Oil Spill Contingency and Response
OSRI   Oil Spill Recovery Institute

PAH   polynuclear aromatic hydrocarbon
PCC   Pollock Conservation Cooperative
PI   principal investigator
PICES   North Pacific Marine Science Organization
POM   Princeton Ocean Model
PWS   Prince William Sound
PWS RCAC Prince William Sound Regional Citizens Advisory Council
PWSSC   Prince William Sound Science Center

R&D	research and development
RAMS	Regional Atmospheric Modeling System
RARE	Resources at Risk
RCAC	Regional Citizens Advisory Council
RFP	Request for Proposals
RRT	regional response team
RSMAS	Rosenstiel School of Marine and Atmospheric Sciences (University of Miami)

SAG	Spill Advisory Group
SCMP	Sustainable Coastal Margins Program
SEA	Sound Ecosystem Assessment program
SeaWiFS	Sea-viewing Wide Field-of-view Sensor (NASA)
SFOS	School of Fisheries and Ocean Sciences
SPM	suspended particulate matter
STC	Scientific and Technical Committee

UAF	University of Alaska, Fairbanks
USCG	U.S. Coast Guard
USFS	USDA Forestry Service